Lecture Notes in Computer Science   5172

Commenced Publication in 1973
Founding and Former Series Editors:
Gerhard Goos, Juris Hartmanis, and Jan van Leeuwen

Editorial Board

T0213139

Shlomi Dolev   Tobias Haist   Mihai Oltean (Eds.)

Optical Supercomputing

First International Workshop, OSC 2008
Vienna, Austria, August 26, 2008
Proceedings

Springer

Volume Editors

Shlomi Dolev
Department of Computer Science
Ben-Gurion University of the Negev
Beer-Sheva, Israel
E-mail: dolev@cs.bgu.ac.il

Tobias Haist
Universität Stuttgart
Stuttgart, Germany
E-mail: haist@ito.uni-stuttgart.de

Mihai Oltean
Computer Science Department
Babeş-Bolyai University
Cluj-Napoca, Romania
E-mail: moltean@cs.ubbcluj.ro

Library of Congress Control Number: Applied for

CR Subject Classification (1998): F.1, B.4.3, C.5.1

LNCS Sublibrary: SL 1 – Theoretical Computer Science and General Issues

ISSN 0302-9743

ISBN 978-3-540-85672-6 Springer Berlin Heidelberg New York

Springer is a part of Springer Science+Business Media

springer.com

© Springer-Verlag Berlin Heidelberg 2008

Typesetting: Camera-ready by author, data conversion by Scientific Publishing Services, Chennai, India
Printed on acid-free paper   SPIN: 12511204   06/3180   5 4 3 2 1 0

Preface

OCS, the International Workshop on Optical SuperComputing, is a new annual forum for research presentations on all facets of optical computing for solving hard computation tasks. Optical computing devices have the potential to build the very next computing infrastructure. Given the frequency limitations and cross-talk phenomena, as well as the soft-errors, of electronic devices on one hand, and the natural parallelism of optical computing devices, as well as the advances in fiber optics and optical switches, on the other hand, optical computing is becoming increasingly marketable. The focus of the workshop is on research surrounding the theory, design, specification, analysis, implementation, and application of optical supercomputers. Topics of interest include, but are not limited to: design of optical computing devices; electro-optics devices for interacting with optical computing devices; practical implementations; analysis of existing devices and case studies; optical and laser switching technologies; applications and algorithms for optical devices; and alpha practical, x-rays and nano-technologies for optical computing. The First OSC workshop was held on August 26th, 2008, in Vienna, Austria, co-located with the 7th International Conference on Unconventional Computing.

This volume contains eight contributions selected by the program committee and two invited papers. All submitted papers were read and evaluated by at least three program committee members, assisted by external reviewers. The review process was aided by the EasyChair system.

OSC 2008 was organized in cooperation with OSA the Optical Society of America. The support of Ben-Gurion University and Babeş-Bolyai University is also gratefully acknowledged.

June 2008

Shlomi Dolev
Tobias Haist
Mihai Oltean

Organization

OSC, the International Workshop on Optical SuperComputing, is a new annual forum for research presentations on all facets of optical computing. OSC 2008 is organized in cooperation with the OSA the Optical Society of America and in conjunction with the International Conference on Unconventional Computation.

Steering Committee

Shlomi Dolev	Ben-Gurion University of the Negev
Mihai Oltean	Babeş-Bolyai University
Wolfgang Osten	Stuttgart University

Organization Committee

Program Chairs	Shlomi Dolev, Ben-Gurion University of the Negev
	Tobias Haist, Stuttgart University
	Mihai Oltean, Babeş-Bolyai University

Program Committee

Hossin Abdeldayem	NASA-Goddard
H. John Caulfield	Fisk University
Shlomi Dolev (Chair)	Ben-Gurion University
Hen Fitoussi	Ben-Gurion University
Joseph W. Goodman	Stanford University
Will Green	IBM T. J. Watson
Lene Hau	Harvard University
Jeremy O'Brien	University of Bristol
Mihai Oltean (Co-chair)	Babeş-Bolyai University
Tobias Haist (Co-chair)	Stuttgart University
Alastair McAulay	Lehigh University
Stephane Messika	LRI
John Reif	Duke University
Joseph Rosen	Ben-Gurion University
Natan T. Shaked	Ben-Gurion University
Damien Woods	University College Cork

Table of Contents

Optics Goes Where No Electronics Can Go: Zero-Energy-Dissipation Logic

H. John Caulfield

Fisk University
1000 17th Ave., N.
Nashville, TN 37208

Abstract. Optical computing has a seemingly eternal problem. It always appears to be in competition with electronic computing. Moore's law and the advantages of digital over analog processing make pure electronics superior in almost every case. Optical computing uses come when the signal is already in the optical domain or when it is used to reduce the heat load in hybrid optical-electronic chips. I describe here work done with a number of bright opticists and logicians over the last four years that produces using optics logic that dissipates no energy and accommodates whatever bandwidth the input and output laser modulation affords. Moreover, we can show why electronics alone can never accomplish those important properties.

Keywords: Optical logic, zero energy, unlimited bandwidth.

1 Introduction

Optical computing has existed for many decades now and has been through many cycles of excitement and depression [1,2,3]. It seems now to be in its best condition ever, because there are niches it seems to fill (optical communication) and next generation chips that will contain optics and electronics. Though these matters are quite important, they will not be reviewed here. Rather, here we report on a totally new capability that optics alone can do. In this, there can never be competition. For the first time, optical computing has a field to itself. That field is sequential logic that does two things that seem at first to be beyond possibility for either electronics or optics:

1. Perform logic with zero energy dissipation.
2. Operate at whatever speed can be modulated onto a laser beam.

The zero energy operation was shown to be thermodynamically allowable many years ago [4,5,6,7]. The reader should note, that the papers cited here are the ones we consider to be the foundation papers. The literature in the field is huge. A limited but useful bibliography of that field has been published online [8].

What Landauer noted [4] was that a binary logic gate takes in two bits but outputs only one. The erasure of one bit costs at least $kT \ln 2$ in energy, where k is the Boltzmann constant and T is the ambient temperature. This is a very

S. Dolev, T. Haist, and M. Oltean (Eds.): OSC 2008, LNCS 5172, pp. 1–8, 2008.

tiny amount of energy. But if we aim for, say, 109 of those per second, which requires a huge amount of power. And current electronic computers use millions of times more energy than this minimum.

So far as I can discern, it was Bennett [5,6] who took the next critical step. Like moist good ideas, it was very simple after he published it. If we do not erase that extra output bit, there is no price to pay for it. They tended to call that undesired bit a "garbage bit." But logic gates so constructed were reversible. Information was transformed but not lost. In a sense nothing was lost. If gates using that strange kind of gate could be made, then lossless logic might appear momentarily.

Fredkin and Toffoli each contributed computationally complete reversible logic gates [7]. This made the growing community of researchers in the field quite excited. Perhaps the end was near. Someone might invent a suitable zero energy or at least a very low energy logic gate soon. They were wrong.

Feynman [9] contributed his own gate and published it in an optics journal. That was really the start of the quest that started me down the road of optical reversible logic. With various colleagues, most notably Joseph Shamir [10], I set out to see if those gate could be made optically. The result was a kind of graceful failure. Later, I will describe precisely the mistake we made. Nevertheless, we accomplished several nice things, only one of which I mention in the introduction, namely: Both Joe Shamir and I became very interested in this field.

In about 2003, I saw an opportunity to assemble a team of bright opticists and bright logicians to try again what we failed to do earlier. Our team (grouped by affiliation) was

- Joseph Shamir, The Technion (consultant)
- Andrey Zavalin, Fisk University (opticist)
- Lei Qian, Fisk University (logician)
- Chandra Vikram, Fisk University (opticist)
- John Caulfield, Fisk University (opticist)
- Jim Hardy, Idaho State (logician)
- Jonathan Westphal, Idaho State (logician)
- Liz Golden, Idaho State (logician)
- Steve Blair, Utah (integrated optics)

Like everyone before us, we struggled but began to see some progress a few years ago. But it was a shock to us when we realized in late 2007 that

1. We had solved the problem
2. We knew why everyone else had failed.

Those are the stories I want to tell in this paper.

2   Zero Energy Dissipation Logic

There were things we knew and things we thought we knew. What we actually knew is shown in Figure 1.

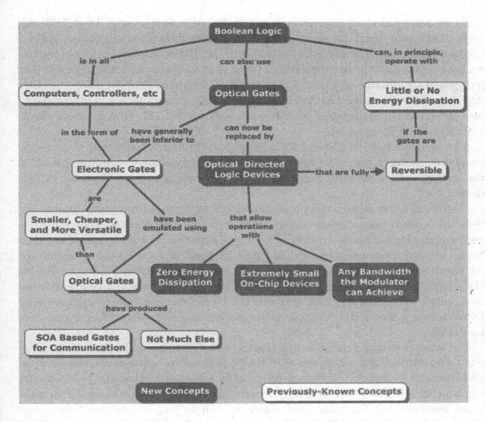

Fig. 1. Our initial insight was that the zero-energy-dissipation requires making the all of the operations linear. That is obvious, but a surprising number of people forget it, for reasons to be discussed below.

A linear operation can be made lossless in information content. A reversible system must conserve information. The arithmetic version of reversibility matter makes it easier to understand. Consider the relationship

$$2+3 = 5$$

Initially, it seems that information has been added. We might not have known that the sum of 2 and 3 is 5. Nevertheless, that is implied by the data (2 and 3) and the instructions (add the numbers). So it really adds nothing that was not already implied. To the contrary, $2 + 3 = 5$ destroys information. Given the answer - 5 - there is no way to find what instructions and data led to that answer. Was it "add 2 and 3"? Perhaps it was add 4 and 1? Perhaps it was: divide 10 by 2. An information conservative operation is

$$2+3 = 3+2.$$

No information is lost, so it could be reversed if we wanted.

So we reasoned that, as there are no lossless gain mechanisms, by definition; each component must be both lossless and reversible. This suggested optics, because optics is inherently linear and reversible. To accomplish nonlinear optics we have to insert some appropriate material that interacts with the light. Of the past papers in optical logic only a very few have sought to do logic using the only way it could conceivably be dome: with only linear components. Those papers were those of my colleagues in this work and me [11,12]. And none of them actually solved the problem.

The originators of this line of thinking [4,5,6,7,8] had in mind only the energy needed to perform the logic operations. Neither they nor we were interested in the energy needed to enter the data not in the energy needed to read out the data.

Historically, we solved the problem twice. Out first solution was limited in that the lossless logic device we produced could be programmed electrically to perform any logical function without energy dissipation. For many purposes, this device (called a Generalized Optical Logic Element or GOLE) is not useful, as its input mechanism was electronic but its output was in the phase of light in the output relative to the phase in another beam. This gave it very limited use and limited cascadability [11,12]. It will not be discussed more extensively here, because it is a wonderful example of an intellectual error we and everyone else had made.

But also, it demonstrated that a passive switch was possible. And that turns out to be vital. Figure 2 shows why that relationship between interferometers and intensity switches. Interferometers can be adjusted so that the two outputs can be either 0 and 1 or 1 and zero, depending on the relative phase of the two input beams. That functions as a binary switch so long as the two beams are present. Electronics and other nonlinear means can latch the output states, while interferometry cannot. This passive, linear switch is the key to our inventions, see Figure 2.

Two things we thought we knew and did not. The most obvious thing about doing reversible logic is that it must be done by linear lossless gate designs such as those of Fredkin, Toffoli, Feynman, and so forth.

The second thing we and everyone else knew that was wrong concerned the relationship of speed to energy loss. Feynman [9] explained the problem with elegant simplicity. Fortunately, he was completely wrong.

The errors made were largely in the implicit assumptions that

- The solution must involve logic gates that are themselves lossless
- The gates must be electronic

The solutions we found violated both of those "obvious" assumptions. We did not use gates of any previously known kind. And, we did not use electronics in computing the answer. Feynman's supposed inescapable speed problem is that the less energy per gate we use, the slower we go. That is based on the need to urge electronic signals along with an electric field. But that is totally irrelevant to optics. Light moves. If it does not move, it is not light. No urging is needed.

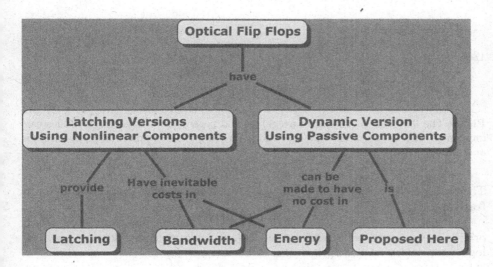

Fig. 2. Conventional logic uses intensity for logic, but that is inherently nonlinear. So it cannot lead to lossless operations. Our optical logic uses interferometry for phase (not intensity) so it can be switched linearly. This clears the path for lossless logic.

Correcting the errors we all made. A major step was taken by Hardy and Shamir [13]. Our earlier work [10] showed clearly why optical Fredkin gates could not produce zero energy logic. The gate has three inputs and three outputs. But one of the inputs and one of the outputs was electronic. The other two inputs and two outputs were optical. But to synthesize even simple logical operations with a Fredkin gate, it must be possible for any output to connect to any input to the next device. With mixed optics and electronics, this cannot work. Hardy and Shamir [13] devised gates that did not require such mixing and thus can be made all optical in terms of the input signals. That matched what optics is good at with the need to evaluate arbitrary sequences of logic operations. What optics does is move (Maxwell's equations require that light travels at the fixed speed c, and Einstein showed that c was the universe's speed limit.) It is easy to direct where the light travels in waveguides, so the need is to figure out how to use the direction of light travel to compute the outcome of a sequence of logical operations. This is what they did in what they called Directed Logic. This is what they did. The results of some simple operations are shown in Figure 3. It gets the right answer, but not in the conventional way.

The earlier assumptions on speed would predict extremely slow operations of any solution to the zero-energy problem. In fact, these gates will work at any speed at which a laser can be modulated. The fastest modulation rates conceivable would be several wavelengths long. The only speed-limiting effect in our devices is clock skew. We need to make all optical path lengths equal; to within a fraction of several wavelengths. This is straightforward to do in optics.

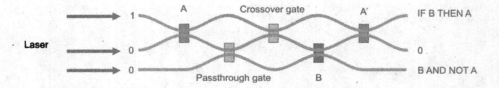

Fig. 3. The flow of light through waveguides and passive interferometric switches can implement any sequence of logic operations. We show one simple but representative operator.

The waveguides cab be made that good, and electrooptic correction is quite feasible.

The last step we have taken is to go back to the regular connection fabric originally suggested by Shamir et al. [10]. It looks nothing like any previous logic gate - electronic or optical. But it works.

Why we (and others) failed for so many decades. Quite often scientists are correct about saying what cannot be done, if their implicit rules apply. We explicitly violated the implicit assumptions of prior workers. Our earlier attempt to solve the problem, explicitly violated the electronics assumption. But we retained the assumption that we must use some sort of constructively complete gates that are themselves lossless.

3   Remaining Problems

Two problems remain.

First, we would like to see how small such systems can be made. At the moment, perhaps tens of microns is required. But, slow light can allow the interaction lengths to be made shorter. If we sacrifice a little on energy, we may be able to use plasmonics to make the devices smaller. If we sacrifice a little in bandwidth, we can use resonant rings that can now be made very small.

Second, we can accomplish the zero-energy feat by avoiding measurements until the end. The system is analog and subject, therefore, to error accumulation. Even here, however, there may be a hope of significant BER (Bit Error Rate). There are two complementary outputs. Measuring both places the decision in the domain of hypothesis testing. This is terra incognito. As we were preparing this paper, we ran across a paper by Peres [14] who uses this same error reduction approach in the kind of optical processor he found so fascinating: those that harness entanglement. This is very encouraging to us, not only because of Peres's great record in his kind of quantum computing but also because his work can be a template for ours.

Summary Diagram. Everything of technical importance in this paper is summarized in Figure 4.

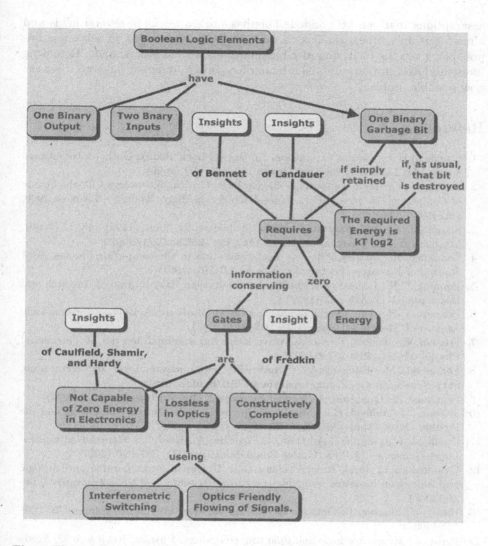

Fig. 4. This is a lossy compression of the history of the project discussed here. Everyone involved made significant contributions but only the lead players on any part of the path toward lossless logic are shown for compactness.

4   Conclusions

The number of implicit assumptions and their persuasive obviousness seem to us what has caused so many brilliant scientists to fail in attempts to make zero-energy-dissipation logic and then be able to use it for high speed operations. We too fell victim to these obvious things for 20 years. We think other "impossible" tasks should be reexamined from time to time to find if they contain implicit assumptions that can be avoided. This has worked for us in several fields and almost certainly others do this as well. The task we set out to solve was impossible, given the tacit, unnoticed assumptions we and others made. Removing inessential assumptions allowed a team of experts in different fields to solve the now-possible problem.

References

1. Caulfield, H.J., Vikram, C., Zavalin, A.: Optical Logic Redux. Optik - International Journal for Light and Electron Optics 117, 199–209 (2006)
2. Shamir, J.: Optics in computing - 40 year later, Critical Technologies for the Future of Computing. In: Proc. SPIE, paper 4109-05 San Diego, 30 July - 4 August 2000, vol. 4109 (2000)
3. Shamir, J.: Optical computing remains in shadow. EETimes (1122) July 17 (2000), http://www.techweb.com/se/directlink.cgi?EET20000717S0072
4. Landauer, R.: Irreversibility and heat generation in the computing process. IBM Journal of Research and Development 5, 183–191 (1961)
5. Bennett, C.H.: Logical reversibility of computation. IBM Journal of Research and Development 17, 525–532 (1973)
6. Bennett, C.H.: The Thermodynamics of Computation - A Review. International Journal of Theoretical Physics 21, 905–940 (1982)
7. Fredkin, E., Toffoli, T.: Conservative logic. International Journal of Theoretical Physics 21(3-4), 219–253 (1982)
8. Perkowski, M.: Bibliography of reversible and quantum logic and computing, http://web.cecs.pdx.edu/~mperkows/PQLG/biblio.html
9. Feynman, R.: Quantum mechanical computers. Optics News 11, 11–20 (1985)
10. Shamir, J., Caulfield, H.J., Miceli, W., Seymour, R.J.: Optical Computing and the Fredkin Gates. Appl. Opt. 25, 1604–1607 (1986)
11. Caulfield, H.J., Soref, R.A., Qian, L., Zavalin, A., Hardy, J.: Generalized Optical Logic Elements - GOLEs. Optics Communications 271, 365–376 (2007)
12. Caulfield, H.J., Soref, R.A., Vikram, C.S.: Universal reconfigurable optical logic with silicon-on-insulator resonant structures. Photonics and Nanostructures 5, 14–20 (2007)
13. Hardy, J., Shamir, J.: Optics inspired logic architecture. Optics Express 15, 150–165 (2007)
14. Peres, A.: Reversible logic and quantum computers. Physical Review A 32, 3266–3276 (1985)

Recent Advances in Photonic Devices for Optical Super Computing

Hossin Abdeldayem[1], Donald O. Frazier[2], William K. Witherow[3], Curtis E. Banks[3], Benjamin G. Penn[3], and Mark S. Paley[4]

[1] NASA Goddard Space Flight Center, Code 305, Greenbelt, MD 20771
[2] NASA Marshall Space Flight Center, Code ED03, Huntsville, Al 35812
[3] NASA Marshall Space Flight Center, Code EV43, ISHM & Sensors Branch,
Huntsville, Al 35812
[4] NASA Marshall Space Flight Center, AZ TECH/Code EM40
Huntsville, AL 35812

1 Introduction

The twentieth century has been the era of semiconductor materials and electronic technology while this millennium is expected to be the age of photonic materials and optical technology. Optical technology has led to countless optical devices that have become indispensable in our daily lives in storage area networks (SANs) [1], parallel processing [2,3], optical switches [4,5], all-optical data networks [6], holographic storage devices [7] and biometric devices at airports [8].

This invited paper is meant to give an overall summary of the latest advances and bring some awareness to the state-of-the-art of optical technologies, which have potential for future super-computing. Optical computing system uses photons instead of electrons to perform appropriate mathematical calculation. In the optical computer of the future, electronic circuits and wires will be replaced by laser diodes, optical fibers, tiny crystals, micro-optical components, and thin films, which will make the systems more efficient, more cost effective, lighter, and more compact. Optical components would not need insulators, as those needed between electronic components, because they are much less sensitive to cross-talk and do not suffer from short circuits. Multiple frequencies of light can travel concurrently through optical components without interference, allowing photonic devices to process multiple streams of data, in parallel, with ease. Optical computing can enhance our computing speed by more than seven orders of magnitude than our current computing speed. This means that an hour of computation by an optical computing system is equivalent to more than eleven years by a conventional electronic computer. Researchers at the University of Rochester have built a simple optical computer, demonstrating the feasibility of such a system, which was able to conduct huge computations nearly instantly. In the last five years, significant advances have been achieved in the field of optical communication to improve upon our communication technology and have a great impact on the development of the optical computing technology. The efforts in trying to avoid the conversion of an optical signal traveling through a fiber to an electronic signal and vise versa and build an all-optical system enhances to a great extent the communication performance and serves very well the optical computing technology. Although recent years have shown a great deal of progress on all fronts in the field of optics,

S. Dolev, T. Haist, and M. Oltean (Eds.): OSC 2008, LNCS 5172, pp. 9–32, 2008.

there are still certain fundamental limitations to be resolved in the optical computing technology, such as cascading, size of optical circuits, integration of components, nonlinear optical processes, laser sizes and powers, etc. Widespread intensive research on the national and international levels is currently progressing at a fast base in academia, industry, and government laboratories to develop the means of processing those light encoded signals without the need for optical conversion to electronic forms. Recent developments in developing all-optical processors, optical switches, optical materials, optical storage media, and optical interconnects have brought all-optical systems closer to reality than ever before as will be shown below. The concept of optical computing stems from the advent of lasers. This promising new technology exploits the advantages of photons over electrons, which include ultrafast information processing and communication. Although an optical computing system is not yet in existence, many related recent developments have been demonstrated which bring the optical computing technology closer to reality. Our intent, in this paper, is to present an overview of the current status of optical computing, and a brief evaluation of the recent advances and performance of the following key components necessary to build an optical computing system.

1. All-Optical Logic gates.
2. Adders
3. Optical processors.
4. Optical Storage
5. Holographic storage.
6. Optical interconnects.
7. Spatial Light Modulators.
8. Optical Materials

2   All-Optical Logic Gates

Logic gates are the building blocks of any digital system. An all-optical logic gate is a switch that controls one light beam by another without the need for an electrical signal; it is "ON" when the device transmits light and it is "OFF" when it blocks the light. We have demonstrated in our laboratory at NASA two ultra-fast all-optical switches in the nanosecond and picosecond regimes using phthalocyanine and polydiacetylene thin films, respectively. The phthalocyanine switch functions as an all-optical AND logic gate, while the polydiacetylene exhibits XOR logic gate functionality as shown in Figure 1.

The AND gate has been demonstrated as follows: the second harmonic from a Nd:YAG laser at 532 nm with a pulse duration of 8 ns collinearly with a probe beam of a cw He-Ne beam at 633 nm have been sent perpendicular to a few hundreds Angstrom thick film of metal-free phthalocyanine. At the output, a narrow band filter has been placed to block the 532 nm beam and allows only the He-Ne beam. The transmitted beam is then focused on a fast photo-detector and sent to a 500 MHz oscilloscope, triggered by the Nd:YAG laser. It has been found that the transmitted He-Ne signal is of a nanosecond duration and is in synchronous with the Nd:YAG pulse. The physical mechanism responsible for this switching effect has been attributed to the saturation (nonlinear) effect, which can easily be explained by assuming that the

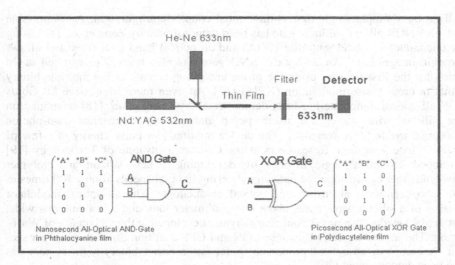

Fig. 1. A schematic of all-optical AND and XOR logic gates, which were demonstrated in phthalocyanine and polydiacetylene thin films in the nanosecond and picosecond regimes, respectively [9]

metal-free phthalocyanine system is a two-level system. The film strongly absorbs the 633 nm from the He-Ne laser as has been observed from its absorption spectrum. When the nanosecond pulse from the green laser, which is also within the film's absorption band, travels through the film, the film becomes saturated (bleached) and allows the red laser to pass through within the duration of the green pulse. After the green pulse is switched off, the system relaxes back to its ground state and blocks the red beam from traveling through. This means that the red output is ON only when both inputs are ON and is OFF when either the green or the red or both are OFF (characteristic table of an AND logic gate). The picosecond XOR gate has been demonstrated in a polydiacetylene thin film, deposited on a substrate that fluoresces red when excited with the green laser. The logic of the XOR gate has been attributed to the excited state absorption, which is also a nonlinear process that has been discussed in detail in reference [10]. The green laser picosecond pulse saturates the first excited state of the polydiacetylene, which strongly absorbs the He-Ne beam and prevents its transmission.

Photonic switches can easily perform in the sub-picosecond (10^{-12}) or femtosecond (10^{-15}) regime as has been demonstrated in polydiacetylene [11]. Li $et.$ $al.$ [12] proposed an all-optical logic gate of SiGe/Si material with multifunctional performance that can function as OR, NOT, NAND, and NOR gates simultaneously or individually. All-optical logic gates based on semiconductor optical amplifiers (SOA) are promising due to their power efficiency and potential for photonic integration [13]. Li $et.$ $al.$[14] proposed a simple and polarization independent logic gate composed of a single SOA followed by an optical band pass filter to achieve various logic functions and have demonstrated AND, OR, and XOR logic functions at 10 Gbit/s. Fujisawa $et.$ $al.$ [15] has proposed a novel design of all-optical logic gates based on nonlinear slot-waveguide couplers, where NOT, OR, and AND logic gates can be realized by a simple single optical-directional coupler configuration. An impressive 40 Gbit/s NOR all-optical logic gate has been demonstrated by Zang *et. al.* [16] using a semiconductor optical amplifier (SOA) and an optical band pass filter and allows photonic integration.   An all-optical AND gate has also been demonstrated at 20 Gbit/s for the first time by using the probe and pump signals as the multiple binary points in cross phase modulation (XPM) [17]. An even more impressive 80 Gbit/s NOR all-optical logic gate has been demonstrated by Liang *et. al.* [18] in submicron size silicon wire waveguide using pump induced non-degenerate two-photon absorption inside the waveguide. The device requires low pulse energy of a few pJ for logic gate operation. Researchers at the California Institute of Technology [19] developed all-optical logic devices by developing a new silicon and polymer waveguide that can manipulate light signals using light, at speeds almost 100 times as fast as conventional electron-based optical modulators. The all-optical modulator consists of a silicon waveguide, about one centimeter long and a few microns wide, that is blanketed with a novel nonlinear polymer developed at the University of Washington. The modulator can be switched ON and OFF a trillion times or more per second. A complete all-optical processing polarization-based binary-logic system has also been demonstrated [20].

3 Adders

An adder is a device which performs the addition of two numbers. Although adders in electronics can be constructed for many numerical representations, the most common adders operate on binary numbers. For single bit adders, there are two general types:

3.1 Half Adder

A half adder (Figure 2) is a logical circuit that performs an addition operation on two binary digits. The half adder produces a sum and a carry value which are both binary digits.

$$S = A \text{ XOR } B$$
$$C = A \text{ AND } B$$

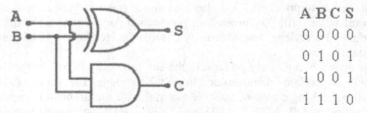

A	B	C	S
0	0	0	0
0	1	0	1
1	0	0	1
1	1	1	0

Fig. 2. Half adder circuit diagram and its logic table

3.2 Full Adder

A full adder (Figure 3) is a logical circuit that performs an addition operation on three binary digits (two inputs and a carry in). The full adder produces a sum and carry value, which are both binary digits. It can be combined with other full adders or work on its own.

$$S = (A\ \text{XOR}\ B)\ \text{XOR}\ Ci$$
$$Co = (A\ \text{AND}\ B)\ \text{OR}\ (Ci\ \text{AND}\ (A\ \text{XOR}\ B)) = (A\ \text{AND}\ B)\ \text{OR}$$
$$(B\ \text{AND}\ Ci)\ \text{OR}\ (Ci\ \text{AND}\ A)$$

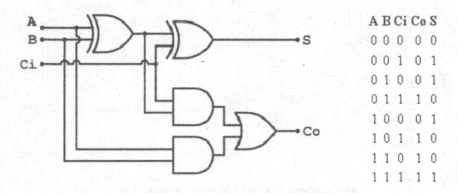

A	B	Ci	Co	S
0	0	0	0	0
0	0	1	0	1
0	1	0	0	1
0	1	1	1	0
1	0	0	0	1
1	0	1	1	0
1	1	0	1	0
1	1	1	1	1

Fig. 3. Full adder circuit diagram: **A + B + Carry In = Sum + Carry Out**

Kim *et.al.* [21] have demonstrated a 10 Gbit/s all-optical half adder using the gain nonlinearity of Cross Gain Modulation (XGM) in semiconductor optical amplifiers. Dong *et. al.* [22] have presented a compact configuration of all-optical adders implemented with a single semiconductor optical amplifier (SOA) and optical band-pass filter (OBF). All-optical adders in photonic bandgap materials containing optically nonlinear layers and using nonlinear optical mechanisms have been demonstrated by several authors in references [23,24], which also lead to other work on the same topic.

4 Optical Processors

A data processor in general is a device that manipulates input information or performs operations on the input data to produce meaningful outputs. As information technology keeps pushing the computing speed and the network capacities towards faster and higher bit rates, the electronic processors at both ends of the fiber-optic communication networks become unable to handle such high data rates. To avoid the inherent cumbersomeness of electronics, all-optical processors will have to eventually replace the electronic ones. Novel processor architectures are essential for future optical computing. A few papers were submitted to this conference suggesting new means of solving NP-Complete problems. Konishi *et.al* [25] described an ultrafast all-optical processor for time-to- 2D-space conversion by using second harmonic generation. They also proposed a technique for the development of an ultrafast all-optical processor that can convert a modulated ultrashort optical pulse sequence into a 2-D spatial distribution for ultrafast spatial information processing using an ultrashort pulse laser. Experimental results show the proposed processor could achieve a throughput of conversion at speeds in the range of Terabits per second (Tbps).

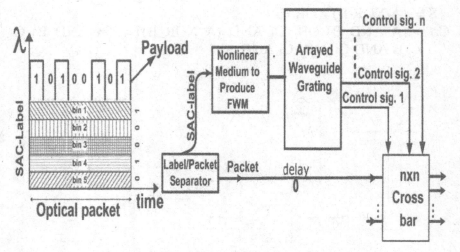

Fig. 4. Schematic of the SAC-label processor [26]

A group of researchers at the University of Laval [26] have proposed a novel label processor, which can recognize multiple spectral-amplitude-code (SAC) labels using the nonlinear optical process of four-wave-mixing (FWM) sidebands and selective optical filtering. In their proposed approach in Figure 4 each label is associated with a spectral amplitude code, which consists of wavelength tones (bins); the payload has been transmitted at a different wavelength. The label recognition unit passes the SAC-label through a nonlinear device to generate frequency sidebands by FWM. If the set of SAC-labels is carefully selected, each code generates at least one unique FWM sideband. The presence of an optical code (OC) can be identified by filtering its unique sideband. This concept is similar to Wavelength Division Multiplexing (WDM) systems that seek to reduce FWM effects by using unequal wavelength spacing [27]. At the output of the nonlinear device, an arrayed waveguide grating acts as the control signal demultiplexer. Splitting of the incoming OC labels is completely avoided and high speed recognition is achieved by the fast nonlinear process. They have succeeded in demonstrating ten SAC-labels on 10 Gbps variable-length packets and achieved error free transmission for more than 5×10^{10} bits in all cases at a bit error rate of less than 2×10^{11} and could transmit over 200 km of fiber with 1dB.

Lenslet has developed a digital signal processor at a new performance level. The processor performs 8 trillion operations per second, which is 1000 times faster than standard processors. It takes multiple electronic digital inputs, converts them into optical signal, performs the desired computation at light speed in the optical core, and then converts the optical output signals back into digital electronic form. It can be used for optical computing, homeland security, military, and multimedia and communications applications [28, 29]. Tian *et.al.* [30] developed an organic material, which can be used as a fast optical processor that compares an input temporal-frequency profile with a recorded reference spectral shape. This pattern –recognition procedure relies on a sub-picosecond temporal cross-correlation process. The size of the phase-encoding spectral interval exceeds 1 THz.

5 Optical Storage

Optical storage is considered nowadays the natural alternative for magnetic recording systems. Optical systems have much more storage capacity and more reliability than magnetic systems. Optical data storage is a general term for all data storage techniques that use optical means to read, write or erase data. Reading all optical storage systems relies on reflected light. Optical compact disc (CD) and Digital Versatile Disc (DVD) drives operate on the principle of detecting the intensity or polarization variations in the optical properties of the media surface. A single disc from Hitachi Maxell, which is only 1 cm longer in diameter than a CD can store up to 300 GB. Currently, a High Definition Digital Versatile Disc (HD-DVD) offers a maximum of 30 GB on a 2-layer disc, and Blu-ray Disc (BD) offers up to 50 GB. The future predictions expect to increase the storage density up to 800GB in two years, and up to 1.6TB by the year 2010. Figure 5 shows a schematic of the different optical storage disks. Hybrid polymer composites, both organic-organic and organic-inorganic are candidate materials for high density storage. New polymers were synthesized as photorefractive information storage materials. Aprilis Inc. demonstrated a composite with a storage capacity of 250 GB on a DVD-like disc with transfer rates exceeding 10 Gbps. The evolution of the different optical storage techniques can be seen in reference [32].

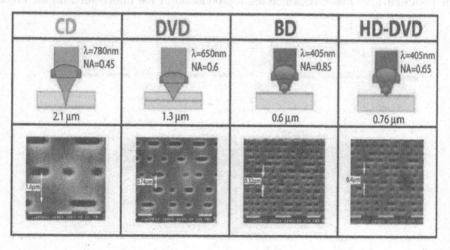

Fig. 5. Schematic of the different optical storage disks, the wavelength, the numerical aperture, and spot size for each [31]

Physicists at Imperial College London [33] in collaboration with other institutions developed a new optical disk the size of a CD or DVD with a storage capacity of one Terabyte. This is ten times more capacity than what a BluRay disk can hold. It is expected to be released between 2010 and 2015. Under magnification the surface of CDs and DVDs appear as tiny grooves filled with pits and land regions, which are encoded as digital formats in the form of ones and zeros. Each bit is read back as one bit. The new technology of Imperial researchers has been designed such that each bit carries ten times the amount of information that we would expect in each bit.   This was achieved by making the bit asymmetric to reflect different information at different orientations.

6 Holographic Storage

The schematic in Figure 6 represents the roadmap of the existing and future technologies for different wavelengths. The surface-storage techniques described above are approaching their fundamental limits.   As the bits become smaller and smaller they become thermally unstable, difficult to access, and unreliable. Holography is a different technique for optically storing information at a much higher density than surface storing techniques because of the capacity for three dimensional imaging. The hologram encodes a large block of data as a single entity in a single write operation and the information is retrieved back as a data block simultaneously. The hologram is constructed as shown in Figure 7 by intersecting two coherent laser beams within a photosensitive material. One of these two beams carries the information to be stored and is called the object or signal beam and the second called the reference beam. The interference grating between the two beams causes chemical or physical changes in the photosensitive material, which are reproduction of the information on the object

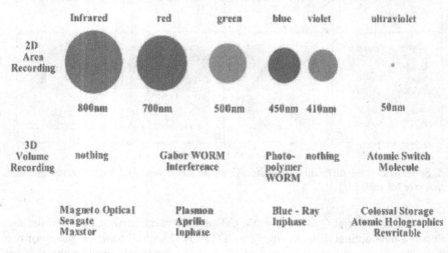

Fig. 6. Schematic of the optical density roadmap[34]

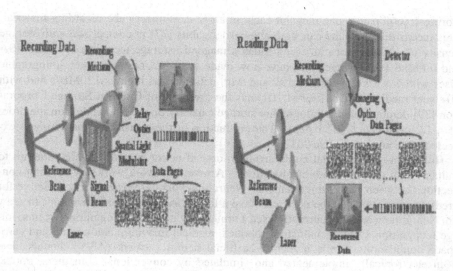

Fig. 7. Schematic of how information is stored and retrieved in a holographic medium (Images courtesy Lucent Technology) [35]

beam. The stored holographic information can be retrieved by illuminating the material by a laser beam at the same wavelength counter propagating to the reference beam. The stored information will be reconstructed along the object beam to be seen by a camera or on a screen.

The key feature of holographic storage is that the data has been stored and retrieved as one page rather than one bit at a time. Multiple pages can be stored simultaneously at the same spot, but at different angles. Up to 500 holograms can be stored in one spot and holograms can be processed at a rate of nearly 1.0 Gbytes / sec. In addition, a defect in the recording medium would not destroy the data bits but rather affect the signal to noise ratio.

Holography breaks the density limitations imposed on us by conventional surface storage techniques. Unlike the encoded digital data stored on conventional CDs, holography allows the writing and reading of billions of data bits with a single flash of light. In volume holography the fundamental storage capacity is limited to ~ V/λ^3, where V is the volume, and λ is the wavelength of the light. This means that holograms enhance both the storage capacity and the rate of data transfer. In addition they are durable, compact, inexpensive, and reliable, which make them the future means of storing large size information and data processing. The storage density of holograms has been achieved by storing many holograms in the same volume at different angles, wavelength multiplexing, orthogonal phase encoding, and fractal–space multiplexing techniques. Wan *et al.* [36] has developed a batch-thermal scheme as a holographic disk storage technique known as the Track-Division Thermal Fixing scheme (TDTF). They have succeeded in storing 5000 images; each contains 768x768 pixels on a disk-shaped 0.03wt% Fe doped LiNbO crystal. Researchers at IBM have succeeded in storing 10,000 pages, one megabit each, in a one centimeter cube recording material. This means that the cube can store about 10 gigabits [37] in comparison with storage density of ~100Kb per cm^2 using today's magnetic storage technique. Maxell USA and InPhase Technologies [38] have now made available, on the market, holographic discs with 1.6 Tera Bytes per disk and with data rates as high as 120MB/s and with 50+ year media archival life. M. Thomas the president of Colossal Storage Corporation [39] demonstrated a holographic storage capacity on a 3.5-inch disc on the order of ~1.2 petabytes (10^{18} bytes). The production of these disks for marketing is expected to be no sooner than 2012.

Holograms are essential components in optical neural network, which attempt to imitate the way a human brain functions. A neural network works by creating connections between processing elements. Neural networks are particularly effective for predicting events when the networks have a large database of prior examples to draw on. Neural networks are currently used prominently in voice recognition systems, image recognition systems, industrial robotics, medical imaging, data mining and aerospace applications. While numerous artificial neural network (ANN) models have been electronically implemented and simulated by conventional computers, optical technology provides a far superior mechanism for the implementation of large-scale ANNs. The properties of light make it an ideal carrier of data signals. With optics, very large and high speed neural network architectures are possible.

7 Optical Interconnects

As optical interconnect technology becomes less costly and better integrated, it also becomes more widely used in links at local and wide area networks, as common alternatives to electrical links. The backbone of the Internet uses fiber optics to transfer data over long distances. Several personal commercial systems such as keyboards, mice, printers, and computers can easily exchange data with each other through invisible infrared light. Several optoelectronic products are also in use to connect rack-to-rack data storage centers, and for network switching. Optoelectronics show future promise as an interconnect technology for exchanging data over ultrashort distances. Data will be exchanged between chips using light instead of electric current, eliminating the need for interconnecting wires. Optical data transfer holds future promise as a more efficient means to move large amounts of data inside a computer rapidly. Multimode optical fibers allow links with a bit rate-distance product of more than 2 GHz.km (i.e., 4 Gb/s over 500 meters), while copper cables have a bit rate-distance product, which is ten times less. Copper cables attenuate electric signals at much faster rates than fiber cables attenuate optical signals. Additionally, optical interconnections do not over-heat as in the case of copper cables. Challenges facing optical interconnections are exacerbated by ultrashort distance for board-to-board and chip-to-chip connections. Intel Corporation has developed an optical interconnect architecture that over the next decade could form the basis for chip-to-chip interconnects, e.g., linkage of a microprocessor to its logic chip set. More information on the status of optical Interconnects can be found in references [40-45].

8 Spatial Light Modulators (SLM)

Spatial Light Modulators (SLMs) are valuable components in building optical computing and neural networks for image processing. SLM modulates either the laser beam intensity or its phase, or both. The most common ones are made of micron-size pixels of Nematic liquid crystals, which have relatively slow modulation speed and a low contrast ratio. This makes them unsuitable for high speed optical computing. Recent advances in the area of SLMs, have been developed using polymeric materials, which have high modulation speed and a high contrast ratio. Holoeye corporation developed high resolution SLMs of 1920x1200 pixels and pixel pitch of 8.1 μm. SLMs (Figure 8) are used extensively in holographic data storage setups to encode information into the laser beam.

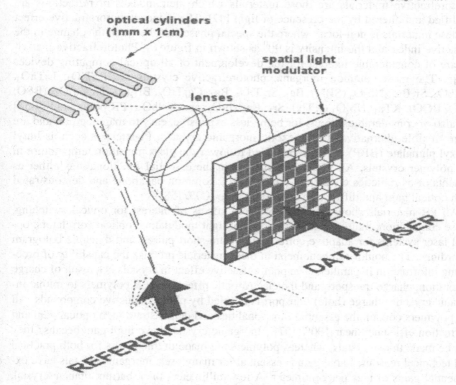

Fig. 8. Schematic of a Spatial Light Modulator setup

SLMs can be electrically or optically addressed. The image on an electrically addressed SLM is created and changed electronically, while the image on an optically addressed SLM is created and changed by shining light encoded with an image on its front or back surface [46] .

9 Optical Materials and New Photonic Devices

Optical materials are crucial elements in the development of all-optical technology. The different materials used to construct the different all-optical components have already been demonstrated but the proper material to build an optical chip that can play the same role as semiconductor chip has not yet emerged. The different materials, which have been examined so far are:

i.   Photorefractive materials
ii. Semiconductors
iii. Doped microshperes
iv. Photonic Band Gap materials
v.   Organic and polymeric materials

i Photorefractive (PR) materials

Photorefractive materials are those materials which their indexes of refractions are modified and altered by the presence of light [47]. The induced photorefractive effect in these materials is non-local, where the special phase shift between the change in the refractive index and the intensity is 90^0 as shown in figure 9. Photorefractive materials are of considerable interest for the development of all-optical computing devices [48]. The most common inorganic photorefractive crystals are $BaTiO_3$, $LiTaO_3$, $LiNbO_3$, $Sr_xBa_{1-x}Nb_2O_6$ (SBN), $Ba_{1-x}Sr_xTiO_3$, $Ba_{1-x}Ca_xTiO_3$, $Bi_{12}(Si, Ti, Ge)O_{20}$ (BSO, BTO, BGO), $KTa_{1-x}Nb_xO_3$ (KTN), $Sr_{1-x}Ba_xNb_2O_6$, and $KNbO_3$. PR polymers [49-56] are also of considerable interest for being less expensive, easy to manufacture, and are more flexible alternatives to the exotic inorganic crystals. Plasticizers such as butyl benzyl phthalate (BBP) [57] can be added to lower the glass transition temperature in PR polymer crystals. All PR polymers contain the essential functionalities either as dopants or as moieties covalently attached to a common backbone and demonstrated high optical gain and diffraction efficiency near 100% [58] .

All PR materials show efficient nonlinear effects of interest for optical switching, phase conjugation, multi-wave mixing, spatial light modulators, optical correlators, optical laser systems for adaptive correction of ultra-short pulses, and dynamic hologram recordings. The nonlinear phenomena in these materials promise the capability of processing information in parallel. The photorefractive effect in crystals is a result of charge generation, charge transport, and the electro-optic effect. For PR polymers to mimic inorganic crystals, charge (hole) transport is enabled by photoconductive compounds. All PR polymers contain the essential functionalities and demonstratd high optical gain and diffraction efficiency near 100% [57]. Inorganic crystals have high-gain because they can be made thick crystals, whereas polymers are made in thin layers for both practical and technical reasons. Large gain is essential for many uses; inorganic crystals have exponential gains of tens per centimeter. A few millimeters thick barium titanate crystal, for example, can show huge net gains, on the order of 10^4 to 10^5. This value means that a 1-mW signal can be amplified to 10 to 100 mW (provided that the other supplied beam has more than 10 to 100 mW). Grunnet-Jepsen *et al.* [58] achieved a net gain of 5 by stacking several layers of their material optically in series while the orientation of the electro-optic PDCST was done electrically in parallel. The gain of 5, although not huge, is sufficient to demonstrate self-pumped phase-conjugation, a particular optical

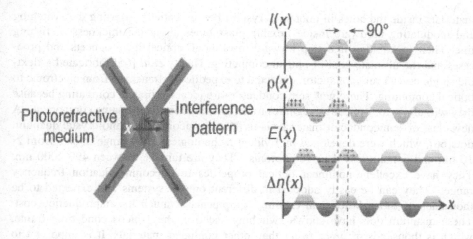

Fig. 9. Hologram formation. (Left) Two beams interfere in a photorefractive medium, producing an index grating across a spatial dimension x. (Right) Relations between spatial light intensity $I(x)$, charge density $\rho(x)$, electric field $E(x)$, and change in index of refraction $Dn(x)$. The $90°$ phase shift between the interference pattern and the index change is indicative of gain, that is, energy transfer from one beam to the other [48].

oscillator configuration used to correct or undo wave-front and image distortions [46]. Breer *et.al.* [59] discussed the wavelength de-multiplexing of superimposed volume-phase holograms in photorefractive Lithium Niobate crystal and demonstrated the de-multiplexing of 1558.0 and 1558.8 nm beams with losses of nearly 15 dB and cross-talk suppression up to 20 dB. Yau *et.al* [60] was able to demonstrate picosecond and femtosecond [61] responses in $BaTiO_3$ crystal by generating a self-pumped phase conjugate signal using a laser of pulse duration 1.5 ps and 126 fs and repetition rates of 86 MHz and 110 MHz, respectively. The possibility of optical limitation of picosecond laser radiation in photorefractive $Bi_{12}SiO_{20}$ (BSO) and $Bi_{12}GeO_{20}$ (BGO) crystals at a wavelength of 1064 nm has been demonstrated [62]. It was shown that the mechanism determining the process of optical limitation is three-photon absorption. In this work, the coefficients of three-photon absorption has been determined by the Z-scanning method.

ii. Semiconductors
Semiconductors are widely used to build lasers of different sizes, and broad range of powers, and wavelengths. These lasers have been used in optical communication, remote sensing, optical data storage, and medical applications. A team from Toronto University, Canada, made a laser by painting a 75 µm diameter glass tube with a solution of nano-sized crystals (quantum dots) of semiconductor lead sulfide [63]. The quantum dots were produced by heating a mixture of oleic acid, lead and sulfur compounds. The glass tube was dipped into the solution and dried afterwards to obtain a thin film of quantum dots. Pumping these quantum dots with a pump source causes them to lase along the tube wall. This laser was useful to connect microprocessors in an optical computer. Sazio et.al.[64] developed a process to embed semi-conducting

materials inside the holes in photonic crystal fiber, essentially inserting the switching and modulating hardware inside flexible glass tubes about the thickness of fishing line. Their technology is a step forward towards all-optical interconnects and proc-esses and the ultimate goal of optical computing. Haetty *et.al* [65] fabricated a flexi-ble single crystal semiconductor, expected to expedite the transition from electronic to optical computing. These new semiconductors are ideal for optical computing because they will allow for optical waveguides to be contained inside the same component. A new class of semiconductor materials is EviDots semiconductor nano-crystal quantum dots[66], which were developed by Evident technologies. They range in size from 2-10 nm and contain less than 1000 atoms. They are tunable between 490-2000 nm. They have excellent nonlinear optical properties in telecommunication frequency range. They can be easily adopted in different optical systems and expected to be thousands of times faster than existing equipments at much lower production cost. These quantum dots have shown switching speed in the 1-picosecond time frame, which is thousands of times faster than other nonlinear materials. It is important to note that all-optical logic gates on an optical chip are unlike electronic transistors on an electronic chip. The electronic transistors on a chip can all be activated by the same power supply, while all-optical logic gates are expected to perform using differ-ent light frequencies and intensities. This is one of the major hurdles in the way of integrating several optical components on a single chip.

iii. Micro-spherical Laser

We recently demonstrated at NASA Goddard Space Flight Center a simple, durable, and inexpensive laser system, which is capable of emitting all sorts of laser lines of interest for an optical chip at the required intensities (patent pending). The system is based on the use of silica microspheres doped with lasing materials of interest and at-tached to a silica fiber. A schematic of the system is shown in Figure 10.

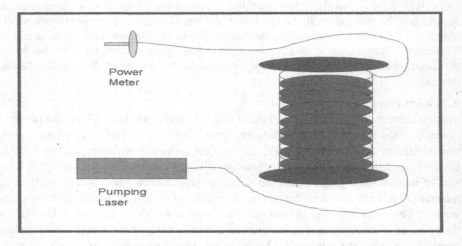

Fig. 10. Schematic of the microspherical-fiber laser system

The coupling of the microspheres to the fiber is achieved as shown in figure 6. The fiber is custom designed where the core is made of silica glass coated with a wax film for protection. The core can be softened at nearly 1500°C and the wax film melts at <90°C. The fiber is coiled around an aluminum spool with its axel attached to a computer controlled motor, not shown. A heating filament on the top is set to a temperature near the softening temperature of the core for a few seconds, where the wax coat is totally melted and the core is exposed. The computer controlled motor rotates the spool for the fine powder spray gun to spray the microspheres at the heated section of the fiber. The process repeats itself for one complete revolution. The spheres and the fiber are then sprayed with an adhesive coating of a refractive index less than that of the spheres and the fiber to stabilize the adherence of the spheres to the fiber. The spool will then be removed away from the heating filament and the spray gun. The cooling element is then activated before turning on the pump laser. The system is contained in a temperature controlled compartment to achieve lasing stability.

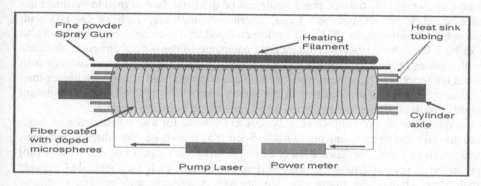

Fig. 11. Schematic of the process of attaching the microspheres to the fiber

iv. Organic Compounds

Organic materials have been extensively investigated for nonlinear optical and electro-optic applications for many years. Thousands of publications about the topic of organic optical materials and their potential for their use in optical devices are presented in scientific journals and conference proceedings. We here present the most recent and interesting ones, which will lead the reader to more materials for gathering further information [67-70].

Organic materials have been demonstrated, as photorefractive materials, optical storage materials, electro-optical materials, and even magnetic materials. These materials can exhibit very high nonlinear optical and electro-optic (EO) coefficients. One such application is display technology. Some concerns with the current state of development of this technology relate, e.g., to space exploration needs, and elevate interest in organic light emitting diodes (OLEDs), an application exploitive of the characteristics of particular organic and polymeric compounds having high EO efficiencies. This particular application would specifically address crew displays, e.g., to be used during certain critical aspects surrounding exploration activity in space. These power-efficient low-mass devices could be good candidates for such operations, while providing an opportunity to expand applications to development of roll-up screens, head wearable displays, and other flexible electronic applications.

Organic Light Emitting Diodes (OLEDs) [71-74] are now at the heart of display technologies breaking into the consumer electronics market and replacing small LCDs found in music players, cameras and mobile phones. Some of the most promising OLED technology materials belong to the class of ionic transition metal complexes (organometallic complexes). These materials have emerged as promising candidates for applications in solid-state electroluminescent devices. The air-tolerant fabrication of such devices and their simple geometry renders them a valuable alternative to the complex multi-layered devices produced from neutral compounds. Four times higher efficiency in OLED devices was demonstrated using organometallic compounds as emitters than using small organic singlet emitter molecules [75]. Furthermore, transition metal complexes can serve as multifunctional chromophores, to address the 3-color stability problem for full-color OLEDs. Using Ru(II)-trisdimine complexes, Rudmann, et. al.[76] showed that ligands can be deliberately designed to optimize device stability and efficiency. Later, alternate complexes, such as Os (II) with phosphine-based ligands [77], and cyclometallated Ir(III) complexes, were shown to exhibit a blue-shifted electroluminescence compared to the 605nm emission observed with devices fabricated with Ru(II) complexes. To date, no single layer devices with ionic luminophores have been reported with emission maxima at energies higher than 560 nm. It is a major objective in the development of useful OLEDs, to pinpoint highly efficient, stable luminophores with blue-shifted emission energies.

A demonstration of conjugated polymers as the basis for a new generation of flexible displays has been promising [77-84]. Scott [85] reports on a conducting polymer, which can be useful for fabrication of basic electronic devices. Shaw and Seidler [86] presented a review article on nontraditional materials such as conjugated organic molecules, short-chain oligomers, longer-chain polymers, and organic–inorganic composites, which can emit light, conduct current, and act as semiconductors. The ability of these materials to transport charge due to the π-orbital overlap of neighboring molecules provides the bases for their semi-conducting and conducting properties. The self-assembling or ordering of these organic and hybrid materials enhances π-orbital overlap and is key to improvements in carrier mobility[87-89]. In addition to their electronic and optical properties, many of these thin-film materials possess good mechanical properties (flexibility and toughness) and can be processed at low temperatures using techniques familiar to the semiconducting and printing industries, such as vacuum evaporation, solution casting, ink-jet printing, and stamping. These properties could lead to new form factors in which roll-to-roll manufacturing could be used to create products such as low- cost information displays on flexible plastic, and logic for smart cards and radio-frequency identification (RFID) tags.

Schon et.al [90] developed an organic solid state injection laser made of a tetracene single crystal using field –effect electrodes for efficient electron and hole injection. Ono et.al.[91] demonstrated a photorefractive polymer-dissolved liquid crystal as a type of intensity amplification of a two-dimensional optical image. A full-color all-optical display was demonstrated [92] using photo induced anisotropy in a bacteriorhodopsin film and a pump-probe beam. The pump beam was the second-harmonic YAG laser and the three probe wavelengths are at 442, 532, and 655 nm from different lasers. PSI-TEC Corporation [93] announced on April, 2006, the successful testing of a nanotechnology engineered organic material called Perkinamine-NR, whose molecular electro-optic coefficient is approximately twice that of the industry's highest- performance materials. Electro-optic materials convert high-speed electronic signals into optical signals and are thus the core active component in high-speed fiber-optics telecommunication, satellite communication, radar, and navigational systems in both the civilian and military sectors. Researchers from Lehigh University and the Swiss Federal Institute of Technology in Zurich (ETH) have reported unprecedented nonlinear optical efficiency in donor-substituted cyanoethynylethene molecules, which can be potentially useful for optical computing, optical data processing, and optical telecommunication[94]. N. Peyghambarian et.al [95] reported that organic optical materials with super-molecular assemblies prepared using chrompohoric building blocks have led to unprecedented electro-optic coefficients greater than 300pm/V as compared to 30 pm/V for lithium noibate.

v. Photonic Band Gap Materials (PBG)

Photonic Band Gap materials are similar to semiconductors, except the electrons are replaced by photons. The current explosion in information technology has been derived from our ability to control the flow of electrons in a semiconductor in some of the most intricate ways. Photonic crystals promise to give us similar control over photons. Given the impact, which semiconductor materials have had on our lives, photonic crystals could play an even greater role in this 21st century, particularly in the optical-communications and optical computing industries [96]. PBG materials are periodic dielectric or metallo-dielectric nano structures with a band gap which forbids certain frequencies of light from passing and can act as a perfect mirror for these frequencies. PBG materials then can be used to control and manipulate the spatial and temporal properties of light propagation. The efficiency of light confinement is limited by the precision with which the structures can be fabricated. The deviation from perfect periodicity of the refractive index reduces the efficiency of light confinement. The Q-factor is a common metric used to evaluate how effectively light is confined within the PBG structure. The Q-factor is also a measure of the rate at which energy escapes from the cavity. The size of the Q-factor is inversely proportional to the cavity size. The nanocavities in PBG materials are important because they can achieve a very high Q-factor and yield high Q optical filter with a wide spectral range of lasers with an ultra low threshold [97]. The Q factor has been expressed as [98]:

$$Q= \lambda/\Delta\lambda$$

Where λ is the center wavelength and $\Delta\lambda$ is the width at half maximum of a resonance. The drawback of a high Q-factor is long photon lifetime (τ) since $Q=\omega\tau$, where ω is optical angular frequency. Significant progress towards practical PBG devices has been accomplished. The highest reported experimental Q-factor of 600,000 was measured for a nanocavity fabricated in a silicon-based two dimensional photonic crystal slab [99]. Photonic crystal microcavity lasers fabricated in InAsP/InGaAsP multi-quantum-well membranes operate with a Q-factor of 13,000 [100]. The reflectance and transmittance of a multilayer structure follows the Bragg condition ($m\lambda=n$ d $\sin\theta$), where m is an integer, λ is the wavelength of the incident light, n the refractive index of the layer, d is the thickness of the layer, and θ is the angle of incidence with respect to the surface.

Yuan et.al. [101] presented a simple and efficient method for computing bandgap structures of a two-dimensional photonic crystal. A great introduction to the topic of photonic bandgap materials can be found in reference [102]. Publications on the topic of photonic bandgap materials have grown exponentially since 1983. These materials hold potential for several applications ranging from optical communications to optical computing as well as inhibition of spontaneous emission to reduce threshold current and noise in semiconductor lasers, low-loss waveguides in optical integrated circuits, perfect mirrors for specific frequencies, optical filters and polarizers. The optical properties of photonic crystal structures can be tuned by modifying its geometrical parameters using thermo-optic effects, carrier plasma effects, electro-optic effects, or by external pressure. Early work suggested that very large refractive index contrasts would be needed to create photonic band gaps in two or three dimensionally periodic photonic crystals. Argyros et.al [103] was able to demonstrate experimentally a photonic band gap fiber made from two glasses with a relative index step of only 1%. Lipson et.al. [104, 105] designed and fabricated a free-space silicon one-dimensional photonic bandgap optical filter. Researchers at NASA's Jet Propulsion Laboratory proposed, in the year 2005, to build high-power fiber lasers with average power levels as high as 1,000 W per fiber made from photonic band gap materials [106]. Researchers at the Department of Energy's Sandia National Laboratory [107] are claiming that they have possibly solved the major technical problem of bending light easily and cheaply without leaking regardless of how many twists or turns are needed for optical communications or optical computing. If this is true, effective use of PBG materials will be a major step forward toward the building of an optical chip, and opens a window to the engineering of dielectric microstructures to make the photons flow in a way similar to electrical currents in semiconductor chips [108].

10 Conclusion

We briefly presented some of the most recent work and components having potential for the manufacture of an optical computing system. The state of the art components demonstrate impressive high speed components with very high storage density and reliability, which brings optical computing closer to reality than ever before. If so, why aren't optical computing systems yet in existence? We believe that there are many hurdles yet to overcome as follows:

1. In the initial development of conventional electronic technology, scientists and engineers availed themselves of the fortuitously existing elements of the periodic table to build the needed transistors. Conversely, optical technology relies, for the most part, on materials yet to be synthesized or developed. Researchers in the field predict that novel materials yet to be derived. Photonic bandgap crystals, are quite promising, and could become the "flesh and bones" of the optical chip for the future optical computer. These materials have the unprecedented ability to guide, rout, control and manipulate light, build laser diodes, optical transistors, confining or even slowing down light, in addition to many other components that can be developed from PBG materials. More importantly, PBGs have the potential for integrating many

components into a single optical circuitry chip. Additionally, such materials offer extremely fast processing speeds, ultra-miniaturization potential, and can be activated by very low power.

2. Electronic transistors on a chip function by the same power supply, while optical components in order to perform, may require different light frequencies and intensities from different laser sources.

3. The issues of miniaturizing optical components, cascading, integrating and processing on a single chip are difficult problems yet to be solved.

4. Optical technology is multidisciplinary and relies on close cooperation between material scientists, physicists, organic chemists, computer architects, computer engineers, computer scientists, and mathematicians. There is a strong need for the government and the industry to integrate and generously fund such inter-disciplined groups.

5. There is still the tendency for scientists and engineers to try imitating the functions of conventional electronic systems when planning and designing an optical computing system. This might not be the optimum way of thinking. There is a need for a paradigm shift in thoughts since electrons and photons are quite different in nature and the means of manipulating them are totally different.

References

1. http://www.rorke.com/med/plasmon-ultra-density-optical.cfm
2. http://e-collection.ethbib.ethz.ch/ecol-pool/diss/fulltext/eth13546.pdf
3. Nitta, K., Matoba, O., Yoshimura, T.: An optical parallel processing for multiplier modulo using an optical interferometer. In: SPIE, August 2006, vol. 6311 (2006)
4. Hamdi, M., Chao, H.J., Blunenthal, D.J., Leonardi, E., Qiao, C., Yun, K.Y.: High-Performance Optical Switches/Routers for High-Speed Internet. IEEE J. on Selected Areas in Communications 21(7) (2003)
5. http://lw.pennnet.com/articles/article_display.cfm?Section=ARTCL&C=Indus&ARTICLE_ID=229349&KEYWORDS=Glimmerglass
6. Hinton, K., Farrell, P., Zalesky, A., Andrew, L., Zukerman, M.: Automatic Laser Shutdown Implications for All-Optical Data Networks. J. of Lightwave Technology 24(2) (2006)
7. Schechter, B., Ross, M.: Leading the Way in Storage, http://domino.watson.ibm.com/comm/wwwr_thinkresearch.nsf/pages/storage297.html
8. http://www.opticsreport.com/content/article.php?article_id=1014
9. Abdeldayem, H., Frazier, D.: Optical Computing:Need and Challenge. Communication of the ACM magazine 50(9) (September 2007)
10. Abdeldayem, H., Paley, M.S., Frazier, D.: An All-Optical Picosecond switch in Polydiacetylene. Applied Physics Letters 82, 1120–1122 (2003)
11. Smith, P.W., Tomlinson, W.J.: IEEE-Spectrum 18, 26 (1981)
12. Li, Z., Chen, Z., Li, B.: Optical pulse controlled all-optical logic gates in SiGe/si multimode interference. Optics Express 13(3), 1033 (2005)

13. Stubkjaer, K.E.: Semiconductor optical amplifier-based all optical gates for high- speed optical processing. IEEE J. Sel. Top. Quantum electron. 6(6), 1428 (2000)

14. Li, Z., Liu, Y., Zhang, S., Ju, H., de Waardt, H., Khoe, G., Dorren, H., Lenstra, D.: All-optical logic gates using semiconductor optical amplifier assisted by optical filter. Electronics Lett. 41(25) (2005)

15. Fujisawa, T., Koshiba, M.: All-optical logic gates based on nonlinear slot-waveguide couplers. J. Opt. Soc. Am. B 23(4), 684 (2006)

16. Zhang, J., Wu, J., Feng, C., Zhou, G., Xu, K., Lin, J.: 40 gbit/s all-optical logic NOR gate based on a semiconductor optical amplifier and a filter. Opt. Eng. 45(8) (2006)

17. Kang, B.K., Kim, J.H., Byun, Y.T., Lee, S., Jhon, Y.M., Woo, D.H., Yang, J.S., Kim, S.H., Park, Y.H., Yu, B.G.: All-Optical AND Gate Using Probe and Pump Signals as the Multiple Binary Points in Cross Phase Modulation. Jpn. J. Appl. Phys. 41, L568–L570 (2002)

18. Liang, T.K., Nunes, L.R., Tsuchiya, M., Abedin, K.S., Miyazaki, T., Thourhout, d.V., dumon, P., Baets, R., Tsang, H.K.: All-optical high speed NOR gate based on two photon absorption in silicon wire waveguides. Optical Soc. Am. Annual Meeting (2005)

19. Scherer, A., Hochberg, M., Baehr-Jones, T.: New All-Optical Modulator Paves the Way to Ultrafast Communications and Computing, http://pr.caltech.edu/media/Press_Releases/PR12901.html

20. Zaghloul, Y.A., Zaghloul, A.R.M.: Complete all-optical processing polarization-based binary logic gates and optical processors. Optics Express 14(21), 9879 (2006)

21. Kim, S., Kim, J., Choi, J., Son, C., Ok, S., Byun, Y., Jhon, Y., Lee, S., Woo, D., Kim, S.: SPIE, vol. 5628, p. 94 (2005)

22. Dong, J., Fu, S., Zhang, X., Shum, P., Huang, D.: All-optical adders based on transient cross phase modulation using a single semiconductor optical amplifier. In: Lee, Y.H., Koyama, F.L. (eds.) SPIE, vol. 6352 (2006)

23. Glushko, E.Y.: All-optical signal processing in photonic structures with shifting bands. Semiconductor Phys., Q. Elec. & Opto-Elec. 7(4), 343–349 (2004)

24. Pei-Li, L., De-xiu, H., Xin-liang, Z., Cuang-xi, Z.: Ultrahigh-speed all-optical half adder based on four wave mixing in semiconductor optical amplifier. Optics Express 14(24), 11839–11847 (2006)

25. Konishi, T., Oshita, Y., Ichioka, Y.: Ultrafast all-optical processor for time-to-two-dimensional space conversion by using second harmonic generation. In: Proc. SPIE, vol. 4110, pp. 182–189 (2000)

26. Rosas–Fernández, J.B., Ayotte, S., LaRochelle, S., Penon, J., Rusc, L.A.: A Single All-Optical Processor for Multiple Spectral Amplitude Code Label Recognition Using Four Wave Mixing, http://www.copl.ulaval.ca/publications/uploadPDF/publication_1864.pdf,
http://www.copl.ulaval.ca/publications/uploadPDF/publication_1865.pdf

27. Forghieri, F., Tkach, R.W., Chraplyvy, A.R.: WDM Systems with Unequally spaced Channels. Journal of Lightwave Technology 13, 889–907 (1995)

28. http://dataweek.co.za/article.aspx?pklArticleId=1678&pklIssueId=46&pklCategoryId=34

29. http://edition.cnn.com/2003/TECH/ptech/10/31/israel.lenslet.reut/index.html

30. Tian, M., Grelet, F., Lorgere, I., Galaup, J.-P., Le Gouet, J.-L.: Persistent spectral hole burning in an organic material for temporal pattern recognition. J. Opt. Soc. Am. -B 16,

74–82   (1999),   http://www.opticsinfobase.org/abstract.cfm?URI=josab-16-1-74

31. Munro, P.R.T., Macias-Romero, Ho, G.-H., Eastley, B., Torok, P.: Optical Data Storage, http://www.imperial.ac.uk/research/photonics/pt_group/peter_torok_research_topics_ODS.htm

32. http://www.opticalstorage.org/article/optical-storage-types/other-types-of-optical-storage.html

33. Torok", P.: 1 Terabyte Optical Storage Disks the Size of a DVD, http://www.physorg.com/news1333.html

34. http://www.physorg.com/news785.html

35. http://computer.howstuffworks.com/holographic-memory2.htm

36. Wan, Y., Wang, Y., Jiang, Z., Liu, G., Wang, D., Tao, S., Edward, C.: High-density non-volatile volume holographic disk storage. In: Society of Photo-Optical Instrumentation Engineers, Bellingham, WA, International (1988) SPIE, Bellingham WA, Etats-Unis (Monographie) (2004), http://cat.inist.fr/?aModele=afficheN&cpsidt=16338253

37. Schechter, B., Ross, M.: Leading the Way in Storage, http://domino.watson.ibm.com/comm/wwwr_thinkresearch.nsf/pages/storage297.html

38. http://ohgizmo.com/2005/11/23/maxell-releases-holographic-torage-medium/, http://www.newlaunches.com/archives/maxell_launches_holographic_storage_16_tb_at_120_mbps.php

39. http://www.p2pnet.net/story/7929http://www.p2pnet.net/story/7929

40. Benner,.A., Ignatowski, M., Kash, J., Kuchta, D., Ritter, M.: Exploitation of optical interconnects in future server architectures. Powers and Packaging 49(4/5) (2005), http://www.research.ibm.com/journal/rd/494/benner.html

41. Berger, C., Beyeler, R., Dangel, R., Dellmann, L., Horst, F., Lamprecht, T., Morf, T., Offrein, B.J., Yamada, F., Hasegawa, M., Numata, H., Taira, Y.: Optical Interconnect Demonstrator with Embedded Waveguides and Butt-Coupled Optoelectronic Modules, in Adaptive Optics: Analysis and Methods/Computational Optical Sensing and Imaging/Information Photonics/Signal Recovery and Synthesis Topical Meetings on CD-ROM, Technical Digest (Optical Society of America, paper IWD2 (2005), http://www.opticsinfobase.org/abstract.cfm?URI=IP-2005-IWD2

42. Louri, A., Weech, B., Neocleous, C.: A Spanning Multichannel Linked Hypercube: A Gradually Scalable Optical Interconnection Network for Massively Parallel Computing. IEEE Transactions on Parallel and Distributed Systems 9(5), 497–512 (1998)

43. Tsai, F.-C.F., O'Brien, C.J., Petrović, N.S., Rakić, A.D.: Analysis of optical channel cross talk for free-space optical interconnects in the presence of higher-order transverse modes. Appl. Opt. 44, 6380–6387 (2005), http://www.opticsinfobase.org/abstract.cfm?URI=ao-44-30-6380

44. Aljada, M., Alameh, K.E., Lee, Y.-T., Chung, I.-S.: High-speed (2.5 Gbps) reconfigurable inter-chip optical interconnects using opto-VLSI processors. Opt. Express 14, 6823–6836 (2006), http://www.opticsinfobase.org/abstract.cfm?URI=oe-14-15-6823

45. Naruse, M., Kawazoe, T., Sangu, S., Kobayashi, K., Ohtsu, M.: Optical interconnects based on optical far- and near-field interactions for high-density data broadcasting. Opt. Express 14, 306–313 (2006), http://www.opticsinfobase.org/abstract.cfm?URI=oe-14-1-306

46. Milewski, G., Engström, D., Bengtsson, J.: Diffractive optical elements designed for highly precise far-field generation in the presence of artifacts typical for pixelated spatial light modulators. Appl. Opt. 46, 95–105 (2007),
http://www.opticsinfobase.org/abstract.cfm?URI=ao-46-1-95
47. http://www.opticsinfobase.org/abstract.cfm?URI=ao-46-1-95
48. Anderson, D.: High Gains for Polymer dynamic Holography". Science 25 277(5325), 530–531 (1997)
49. Günter, P., Huignard, J.P.: Photorefractive Materials and Their Applications, vol. 1&2. Springer, Berlin (1988 & 1989)
50. Ducharme, S., Scott, J.C., Twieg, R.J., Moerner, W.E.: Observation of the photorefractive effect in a polymer. Phys. Rev. Lett. 66, 1846–1849 (1991)
51. Tamura, K., Padias, A.B., Hall Jr, H.K., Peyghambarian, N.: New polymeric material containing the tricyanovinylcarbazole group for photorefractive applications. Appl. Phys. Lett. 60, 1803–1805 (1992)
52. Cui, Y., Zhang, Y., Prasad, P.N., Schildkraut, J.S., Williams, D.: Photorefractive effect in a new organic system of doped nonlinear polymer. J. Appl. Phys. Lett. 61, 2132–2134 (1992)
53. Kippelen, B., Tamura, K., Peyghambarian, N., Padias, A.B., Hall Jr., H.K.: Phys. Rev. B 48, 10710–10717 (1992)
54. Donkers, M.C.J.M., Silence, S.M., Walsh, C.A., Hache, F., Burland, D.M., Moerner, W.E., Twieg, R.J.: Net two-beam-coupling gain in a plymeric photrefravtive material. Opt. Lett. 18, 1044–1046 (1993)
55. Sandalphon, B.K., Peyghambarian, N., Lyon, S.L., Padias, A.B., Hall Jr., H.K.: New Highly efficient photrefractive polymer composite for optical-storage and image processing applications. Electron. Lett. 29, 1873–1874 (1993)
56. Liphard, M., Goonesekera, A., Jones, B.E., Ducharme, S., Takacs, J.M., Zhang, L.: High-performance Photorefractive polymers. Science 263, 367–369 (1994)
57. Meerholz, K., Volodin, B.L., Sandalphon, B.K., Peyghambarian, N.: A photorefractive polymer with high optical gain and diffraction efficiency near 100%. Nature 371, 497 (1994)
58. Grunnet-Jepesen, A., Thompson, C.L., Moerner, W.E.: Spontaneous Oscillation and Self-Pumped Phase conjugation in a Photorefractive Polymer Optical amplifier. Science 277, 549 (1997)
59. Breer, S., Buse, K.: Wavelength demultiplexing with volume phase holograms in photorefractive lithium Niobate. Appl. Phys. B. 66(3), 339 (2004)
60. Yau, H., Pan, E., Wang, P., Chen, J.: Phase Conjugation with Picosecond Pulses in Ba-TiO3. Opt. Rev. 3(5), 312 (1996)
61. Yau, H., Pan, E., Wang, P., Chang, C., Cheng, N., Chen, J.: Mechanism for Ultra-Short Phase conjugate Pulse with Photorefractive Crystal. Chinese Journal of Physics 36(6), 791 (1998)
62. Ryasnyanskii, A.I.: Three-Photon Absorption in Photorefractive BSO and BGO Crystals. J. Appl. Spectr. 71(2), 295 (2004)
63. Hoogland, S., Sukhovatkin, V., Howard, I., Cauchi, S., Levina, L., Sargent, E.: A solution-processed 1.53 mm quantum dot laser with temperature-invariant emission wavelength. Optics Express 14(8), 3273 (2006),
http://www.newscientisttech.com/article/dn9017,
http://www.news.utoronto.ca/bin6/060419-2208.asp
64. Sazio, P.A., Amezcua-Correa, A., Finlayson, C.E., Hayes, J.R., Scheidemantel, T.J., Baril, N.F., Jackson, B.R., Won, D.J., Zhang, F., Margine, E.R., Gopalan, V., Crespi, V.H., Badding, J.V.: Microstructured Optical Fibers as High-Pressure Microfluidic Reactors. Science 311(5767), 1583 (2006),
http://www.sciencemag.org/cgi/content/full/311/5767/   1583,
http://www.physorg.com/news81621455.html

65. Haetty, J., Na, M.H., Chang, H.C., Luo, H., Petrou, A.: Fabrication of flexible monocrystalline ZnSe based foils and membranes. Appl. Phys. Lett. 69(11), 1608 (1996), http://www.buffalo.edu/reporter/vol28/vol28n12/f2.html, http://www.scienceblog.com/community/older/1996/A/199600445.html
66. http://www.oceanoptics.com/products/evidots.asp
67. Peyghambarian, N., Norwood, R.A.: Organic Optoelectronics Materials and Devices for Photonic Applications, Part One. Optics & Photonics News 16, 30-35 (2005), http://www.opticsinfobase.org/abstract.cfm?URI=OPN-16-2-30
68. Peyghambarian, N.P., Norwood, R.A.: Organic Optoelectronics Materials and devices for photonic applications, part II. Optics and Photonics News 16(4), 28 (2005)
69. Jen, A., Luo, J., Kim, T., et al.: Proc. SPIE 5935, 5935061 (2005)
70. Li, J., Hu, L., Wang, L., Zhou, Y., Grüner, G., Marks, T.J.: Organic Light-Emitting Diodes Having Carbon Nanotube Anodes. Nano Lett. 6, 2472–2477 (2006)
71. Luo, J., Haller, M., Ma, H., et al.: J. Phys. Chem. B 108(25), 8523 (2004)
72. Dalton, B.R., Jen, A., et al.: Proc. SPIE 5935, 5935021 (2005)
73. Shinar, J.: Organic Light Emitting Diodes. Springer, New York (2004)
74. Li, Y.Q., Rizzo, A., Cingolani, R., Gigli, G.: Bright White-Light-Emitting Device from Ternary Nanocrystal Composites. Advanced Materials 18(19), 2545 (2006)
75. Kalinowski, J.: Organic Light Emitting Diodes: Principles, Characteristics and Processes. Marcel Dekker, New York (2005)
76. Rudmann, H., Shimada, S., Rubner, M.F.: High-efficiencer light-emitting devices based on derivatives of the tris(2,2'-bipyridyl)ruthenium(II) complex. J. Am. Chem. Soc. 124, 4918 (2002)
77. Bernhard, S., Malliaras, G.G., Abru-a, H.D.: Advanced Materials. Efficient Electroluminescent Devices Based on a Chelated Osmium(II) Complex 14, 433–435 (2002)
78. Kim, Y.C., Cho, S.H., Song, Y.W., Lee, Y.J., Lee, Y.H., Do, Y.R.: Planarized SiNx/spin-on-glass photonic crystal organic light-emitting diodes. Appl. Phys. Lett. 89, 173502–173504 (2006)
79. Friend, R.H., Gymer, R.W., Holmes, A.B., Burroughes, J.H., Marks, R.N., Taliani, C., Bradley, D.D.C., dos Santos, D.A., Brédas, J.L., Lögdlund, M., Salaneck, W.R.: Conjugated Polymer Electroluminescence. Nature 397, 121–128 (1999)
80. Brédas, J.L., Beljonne, D., Coropceanu, V., Cornil, J.: Charge-Transfer and Energy-Transfer Processes in π-Conjugated Oligomers and Polymers. Chemical Reviews 104, 4971–5004 (2004)
81. Brédas, J.L., Calbert, J.P., da Silva Filho, D.A., Cornil, J.: Organic Semiconductors: A Theoretical Characterization of the Basic Parameters Governing Charge Transport. Proceedings of the National Academy of Sciences USA 99, 5804–5809 (2002)
82. Coropceanu, V., Malagoli, M., da Silva Filho, D.A., Gruhn, N.E., Bill, T.G., Brédas, J.L.: Hole- and Electron-Vibrational Couplings in Oligoacene Crystals: Intramolecular Contributions. Physical Review Letters 89, 275503 (2002)
83. Coropceanu, V., André, J.M., Malagoli, M., Brédas, J.L.: The Role of Vibronic Interactions on Intra- and Inter-Molecular Electron Transfer in π-Conjugated Oligomers. Theoretical Chemistry Accounts 110, 59–69 (2003)
84. Malagoli, M., Coropceanu, V., da Silva Filho, D.A., Brédas, J.L.: Multimode Analysis of the Gas-Phase Photoelectron Spectra in Oligoacenes. Journal of Chemical Physics 120, 7490–7496 (2004)
85. Campbell Scott, J.: Conducting polymers:from Novel Science to New Technology. Science 278(5346), 2071–2071 (1997)

86. Shaw, J.M., Seidler, P.F.: Organic Electronics:Intoduction. IBM Journal of Research and Development 45(1) (2001)
87. Dimitrakopoulos, C.D., Malenfant, P.R.L.: Organic Thin Film Transistors for Large Area Electronics. Adv. Mater. 14(2) January 16 (2002)
88. Forrest, S.R.: The path to ubiquitous and low-cost organic electronic appliances on plastic. Nature 428, 911–918 (2004)
89. Reese, C., Roberts, M., Ling, M., Bao, Z.: Organic thin film transistors. Materials Today 7(9), 20–27 (2004)
90. Schon, J.H., Kloc, C., Dodabalapur, A., Batlogg, B.: An Organic Solid State Injection Laser. Science 289(5479), 59–601 (2000)
91. Ono, H., Emoto, A., Kawatsuki, N.: Reconstruction of two-dimensional optical image from nonlocal gratings in a photorefractive mesogenic composite. Optical Materials 27(3), 509–514 (2004)
92. Huang, Y., Siganakis, G., Moharam, M.G.J., Wu, S.-T.: All-optical display using photoinduced anisotropy in a bacteriorhodopsin film. Opt. Lett. 29, 1933–1935 (2004)
93. http://www.nanotechwire.com/news.asp?nid=3199&ntid=122&pg=2
94. May, J.C., Lim, J.H., Biaggio, I., Moonen, N.N.P., Michinobu, T., Diederich, F.: Highly Efficient Third-Order Optical Nonlinearities in Donor-Substituted Cyanoethynylethene Molecules. Opt. Lett. 30, 3057 (2005)
95. Peyghambarian, N., Dalton, L., Jen, A., Kippelen, B., Marder, S., Norwood, R., Perry, J.W.: NONLINEAR OPTICS: Technological advances brighten horizons for organic nonlinear optics. Laser Focus World, 85–92 (August 2005)
96. http://physicsweb.org/articles/world/13/8/9
97. Baba, T.: Remember the light. Nature 1(1), 11–12 (2007)
98. Weiss, S.M.: Tunable Porous Silicon Photonic Bandgap Structures: Mirrors for Optical Interconnects and Optical Switching. Ph. D. thesis, the institute of optics, school of engineering and applied sciences, University of Rochester (2005)
99. Song, B.S., Noda, S., Asano, T., Akanane, Y.: Ultra-high-Q photonic double-heterostructure nanocavity. Nature Materials 4, 207–210 (2005)
100. Srinivasan, K., Barclay, P.e., Painter, O., Chen, J., Cho, A.Y., Gmachl, C.: Experimental demonstration of a high qulity factor photonic crystal microcavity. Appl. Phys. Lett. 83, 1915–1917 (2003)
101. Yuan, J., Lu, Y.Y.: Photonic bandga caluculations with Dirichlet-t Neumann maps. J. Opt. Soc. Am. A 23, 3217–3222 (2006),
http://www.opticsinfobase.org/abstract.cfm?URI=josaa-23-12-3217,
http://www.elektrorevue.cz/clanky/2004/0003,
http://www.opticsinfobase.org/abstract.cfm?URI=josaa-23-12-3217
102. http://www.elettra.trieste.it/experiments/beamlines/lilit/ht docs/people/luca/tesihtml/node3.html
103. Argyros, A., Birks, T., Leon-Saval, S., Cordeiro, C.M., Luan, F., Russell, P.S.J.: Photonic bandgap with an index step of one percent. Optics Express 13(1), 309–314 (2005)
104. Lipson, A., Yeatman, E.: Low-loss one-dimensional photonic band gap filter in (110) silicon. Opt. Lett. 31, 395–397 (2006)
105. Yeatman, E.M., Lipson, A.: Silicon MEMS for photonic bandgap devices. In: Proc. NSTI Nanotech 2006, Boston, May 7-11, pp. 409–412 (2006)
106. http://www.techbriefs.com/content/view/152/34
107. http://www.sandia.gov/media/photonic.htm
108. Busch, K., John, S.: Liquid-crystal photonic bandgap materials: the tunable electromagnetic vacuum. Phys. Rev. Lett. 83, 967–970 (1999),
http://focus.aps.org/story/v4/st7

Ultrafast Digital-Optical Arithmetic Using Wave-Optical Computing

Tobias Haist and Wolfgang Osten

Institut für Technische Optik
Universität Stuttgart
Pfaffenwaldring 9, 70569 Stuttgart, Germany
haist@ito.uni-stuttgart.de

Abstract. We propose a method for implementing digital-optical arithmetic with high accuracy at extremely high speed. To this end we use the superposition of photons running through a passive network of simple optical components. All possible solution are realized in parallel by superposition. Therefore, the overall computing time can be reduced to the sum of the time of flight through a very short optical path and the time needed for input and output.

1 Introduction

Fig. 1 shows the classical approach for performing computations. Input signals are processed before we obtain the results. Obviously the overall time for performing the computation is the sum of the time needed for input/output, the time for the computation itself, and the time for the signals to travel from input to output.

In the early days of optical computing the speed of the proposed methods typically was limited completely by the switching time of light- or electronically-controlled switches and detectors. For very fast switching we need reasonable large power because always a minimum switching energy is required. Also detection requires a minimum number of photons, and therefore detection time somehow is anti-proportional to the available energy. In this paper we neglect such energy considerations. We are not interested in an energy efficient, but rather in the fastest possible system.

During the last two decades — mostly due to the tremendous progress in telecommunications — a lot of research effort went into the development of detectors and switches. Switching and detection of photons with femtosecond speed is possible[1, 2], and with such switches a lot of processing methods are limited by the time of flight of the photons running through the optical paths. Therefore, tomorrow's final speed limitation for optical computing architectures is the optical path length that the photons have to travel. According to Fig. 1 (a) this means that the overall propagation distance should be as small as possible.

Instead of trying to minimize current optical processors we propose to use the system of Fig. 1 (b). At first one might think that this is impossible, because the input is introduced *after* the processing stage.

Fig. 1 (b) might·be regarded as a simple look-up approach of precomputed results, that is employing the well known technique of trading memory for speed.

S. Dolev, T. Haist, and M. Oltean (Eds.): OSC 2008, LNCS 5172, pp. 33–45, 2008.

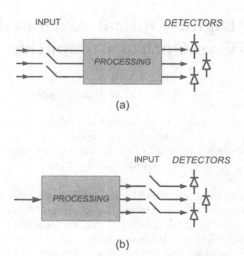

Fig. 1. Computing as it is done (a) normally and (b) as we propose it

But this is not exactly what we propose here. Although such a conventional look-up approach is powerful, it is only applicable if the number of possible input values is limited. As an example take the digital addition of two 32 bit numbers. A simple look-up table would need $2 \cdot 32 = 64$ bits of input. Therefore, we would need a table with 2^{64} entries. Of course, this would not be feasible.

In the following, we will propose a method for using the scheme of Fig. 1 (b) without any precomputed and stored results. The core idea is to use the superposition of all possible computations by coding numbers via different path lengths. By that approach, the computation time — defined here as the time between switching the input and reading the output — will be as small as the time of flight between switch and detector. As we will show in the course of this paper, the detector can be located immediately behind the switch, therefore, this time of flight can in principle be made extremely small, so that computing in the femtoseconds region should be possible.

In Sect. 2 we introduce the core of wave-optical digital computing. The described optical setups are not really practical, but serve to explain the idea. Based on a simple analysis, we extend the method in Sect. 3 to residue arithmetic which fully exploits the maximum achievable speed while enabling high accuracy by parallelization. In Sect. 4 and 5 we describe one possible optical implementation and discuss how the method is related to quantum computing before ending with a short conclusion.

2   Proposed Method

A lot of methods have been proposed and used in the past for the digital-optical implementation of arithmetic operations[3–14]. For a good introduction we recommend the paper of McAulay [15] or the review of Sawchuk[16]. These methods typically make use of polarisation, phase, intensity, color, or spatial position for digitally coding numbers.

Here, we propose to use the optical path length for coding. At first this might seem counterproductive, because increasing the path length leads to delay and therefore, a reduction of the maximum achievable speed. However, we will see that this is not necessarily true. Using path lengths together with white light interferometry[17] allows us to employ the superposition of all possible computations. That means, we use the fact that a photon running through a system of splitted channels simultaneously runs through all possible paths. So, we exploit the inherent parallelism when guiding a large number of photons through an optical network.

Coding solutions by paths of photons has been described already in [18–23] for solving the Hamiltonian path problem and the well-known travelling salesman problem. Here, we extend this principle to digital-optical computations. Several optical realizations of quantum computing that use different paths have been proposed in order to optically realize quantum computing algorithms. Therefore, there is also some kind of relatedness to quantum computing (compare Sect. 5).

Fig. 2 shows a simple example for the analog optical addition of two numbers A and B. It is basically a white light interferometric setup with coherence length L. The reference path length is C. We will obtain interference (constructive or destructive, based upon the chosen alignment) at the detectors if and only if $C = A + B$. Therefore, we might move the mirrors in the lower path until we detect interference. The coherence length of the source determines the accuracy of this analog optical method.

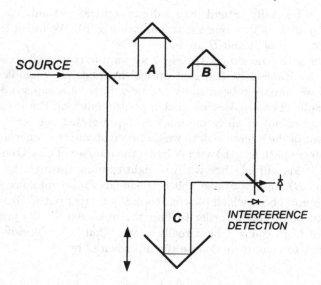

Fig. 2. Analog optical addition of A and B. Interference will be detected if $C = A + B$.

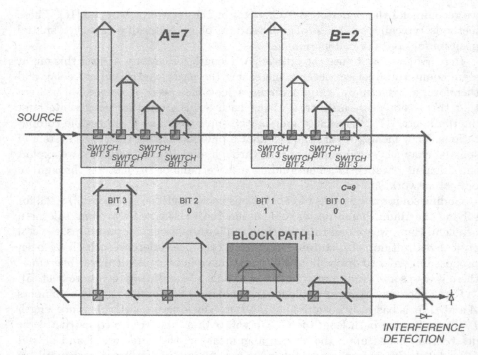

Fig. 3. Digital-optical addition $C = A + B$ in binary representation. The detection of bit #1 of the solution is depicted. Slow version.

This idea can be easily extended to a digital-optical method. We implement A, B, and C by their binary representation (see Fig. 3). We might use switches to control which bits of A and B are set.

For the lower path (encoding C) in Fig. 3 we want to realize all possible integer path lengths simultaneously. Different optical methods are possible to achieve this. In Fig. 3 we employed beamsplitters (for a different setup see Sect. 4). We use the beamsplitters to make sure that a photon entering the lower path will run exactly once or not at all through a loop representing one bit.

At the output of the reference path we therefore obtain the superposition of all 2^N possible integer path lengths with N being the number of bits. One of these 2^N outgoing light fields will interfere with the light running through the upper path (through A and B), and, of course, this corresponds to the unknown solution C.

How can we now check which photons took the correct path? Fortunately, we do not have to know this in order to find the unknown C. We just block the loop of bit i of C to check if bit i really is set. That is by blocking loop i we artificially set c_i to zero, if we denote the unknown C by

$$C = \sum_{0}^{N-1} c_j 2^j \ . \tag{1}$$

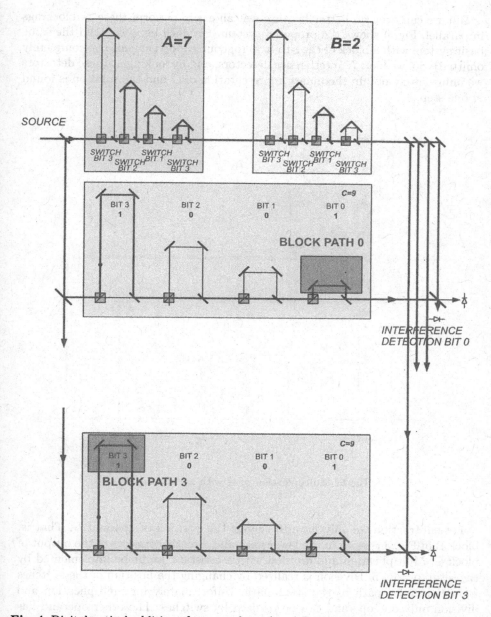

Fig. 4. Digital-optical addition of two numbers A and B in parallel but still slow. Only two channels (bit 0 and bit 1) are depicted.

If there is still interference at the detector, we know for sure that indeed bit i of the solution equals zero. Otherwise we have to conclude that bit i equals one. So, by introducing the N blocks sequentially we might measure the N bits of C.

But we can even do better because we can easily perform these N blockings in parallel. Fig. 4 shows the proposed system. For each bit i we build the same basic system with a block of the i-th loop (of course loop i can also be completely omitted). So, we have N interference detectors, and by looking at these detectors we immediately obtain the binary representation of C and the solution is found in one step.

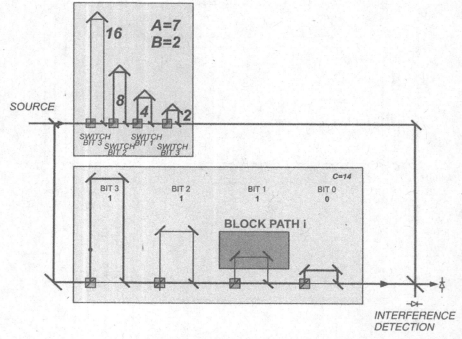

Fig. 5. Multiplication of A with a constant B

For subtraction we only have to change the position of operand B. That is, block B of Fig. 3 goes to to the lower path and directly connects to the output of block C. Multiplication and division with a constant might be implemented by the setup of Fig. 5. Division is realized by changing the position of the switches from the upper path to the lower path. Unfortunately for multiplication and division only one operand can be realized by switches. The other operand has to be hard-coded by choosing the appropriate delay lines.

How fast is such a computation? At first, one might assume that the speed is completely determined by the time a photon needs to travel through the network before interfering at the detector. Fortunately, this is not true. When switching one bit of B, it takes only the time that the photon travels until it hits the detector *after* passing the switch (provided that there are already photons in the network). In a simple classical model this is quite obvious.

In the quantum model things are not so simple, but the famous delayed-choice experiment[24] showed that indeed interference will still persist if the photon already travelled through all possible paths of C *before* we decide how to setup A and B. The crux of this is that we should build our system in such a way that the distance between detector and switches is minimized. By making this distance very small, we can compute complex operations as fast as the time of flight between the switches and the detector. This is in contrast to other approaches of digital-optical computation and only works due to the superposition of all possible results.

3   Residue Arithmetic

We discussed that in order to maximize speed we have to minimize the distance of the switches to the detector. We denote the coherence length of our light source by L which acts as the final quantization of the method. For an N bit number it then takes at least $2^{N-1} \cdot L/c$ until the photon arrives at the detector after switching the most significant bit. For low accuracy this results in reasonable values, but suppose that we want to achieve an accuracy of 32 bit. In this case the system would be extremely slow.

We can improve by giving up the binary representation and having one separate optical path (and one switch) for every number. Fig. 6 shows an example for subtraction. Now, we can put the detectors directly after the switches, and therefore, we obtain ultrafast operation. Unfortunately, this only works for low accuracy. Suppose that we want to achieve 32 bit accuracy. In this case we would need 2^{32} paths and switches which is, of course, not practical.

By using residue arithmetic[5] it is possible to obtain high accuracy together with high speed. Residue arithmetic is a well known technique, dating back at least 1500 years, for representing large integer numbers by multiple small integers. It allows to perform addition, subtraction, multiplication, and polynomial evaluation without carry operations. This means that we can perform arithmetic operations with high accuracy in one single step. Due to this advantage it has been proposed for electronic as well as digital-optical computation[5, 4, 25–27, 16].

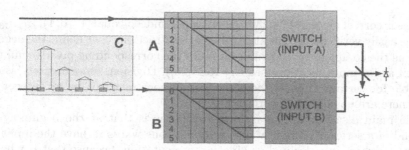

Fig. 6. Ultrafast subtraction based on a decimal representation of the numbers. The detectors can be located directly after the switches.

The basic principle is to represent an integer I by a set of M residues $(R_{N1}, R_{N2}, ..R_{NM})$. A residue R_{Ni} is the remainder of I divided by Ni, that is

$$R_{Ni} = I \bmod Ni := I \% Ni .\qquad(2)$$

Therefore $0 <= R_{Ni} < Ni$.

A simple example explains this more clearly: If we want to represent the integer 1817 in the residue system given by $N1 = 17$, $N2 = 13$, and $N3 = 11$, we obtain

$$R_{N1} = 1817 \% 17 = 15$$
$$R_{N2} = 1817 \% 13 = 10$$
$$R_{N3} = 1817 \% 11 = 2$$

Therefore, $1817 = (15, 10, 2)_{res(17,13,11)}$.

The maximum number that we can represent in the residue system is given by the product of the Ni:

$$I_{max} = \prod_{i=1}^{M} Ni \qquad(3)$$

In order to unambiguously use the number system defined by the Ni, it is required that the Ni are "relatively prime". That means that for every pair (Ni, Nj) the Ni and Nj have no prime factors in common. If we chose prime numbers for the Ni, we easily fulfil this requirement.

As already mentioned, the advantage of this rather complicated representation is that addition, subtraction, multiplication, and polynomial evaluation[5] are carry-free. That is, we can perform all operations on the R_{ni} in parallel.

Again, we use an example to clarify this. The addition of $I_1 = 1817$ and $I_2 = 230$ in the system $(17, 13, 11)$ results in

$$
\begin{array}{rccc}
I_1 = & (15 , & 10 , & 2) \\
I_2 = & (9 , & 9 , & 10) \\
\hline
I = & (7 , & 6 , & 1)
\end{array}
$$

This is correct since $1817 + 230 = 2047$ which indeed equals $(7, 6, 1)_{res(17,13,11)}$.

One easily verifies that this also works for subtraction (if we define the negative value as the complement according to Ni of the corresponding positive number) as well as for multiplication (see e.g. Ref. [25]). Division, unfortunately, is only possible for certain dividers, because the quotient may not be an integer and also there are no sign and overflow bits.

The main disadvantage of residue arithmetic is that at the beginning and the end of a series of computations, normally, one wants to have the input and output numbers in a decimal or binary representation, because that is what we normally are used to work with. The decimal-to-residue conversion (Eq. (2)) can be done even optically[4] and for the residue-to-decimal conversion we can use the Chinese Remainder Theorem or the Mixed Radix Technique[28]. Of course, such conversions will take time, but first, for a longer sequence of operations this is negligible, and second, we are anyway not forced to use the decimal system.

4   One Possible Implementation and Practicability

We are now in the position to combine the core idea of Sect. 2 with residue arithmetic. A lot of practical implementations are possible, we don't want to discuss the pros and cons of them here, and we don't discuss details (e.g. dispersion vs. coherence length) of such implementations. We instead will show a conceptually simple scheme.

Suppose we want to achieve 24 bit accuracy. We might take the four largest prime numbers below 100, that is $N1 = 79$, $N2 = 83$, $N3 = 89$, and $N4 = 97$ as the basis for our residues. According to (3) this results in $I_{max} = 56.606.581$ which corresponds to an accuracy of more than 25 bit.

Fig. 7 shows one possible realization for detecting one bit of one residue for a 7-digits residue base. For the detection of the interference different methods are possible. We schematically have shown here two photodetectors. Due to the π phase shift at the beamsplitter one of the detectors will detect negative interference if the other detector detects positive interference. Other possibilities are switches based on four-wave mixing[29]. In Fig. 7 the different path lengths are realized by continued splitting with blazed gratings. Of course, we have to use similar devices for all the other bits. Addition can be achieved by subtraction of A and the (residue-)complement of B which we easily can implement by relabeling the switches of B. The extension to multiplication with fixed numbers (compare Fig. 5) is straight forward.

For the system with (79, 83, 89, 97) and an addition of two 24 bit numbers, seven bits per residue are sufficient. Therefore, we need 4 x 7 = 28 detection bits

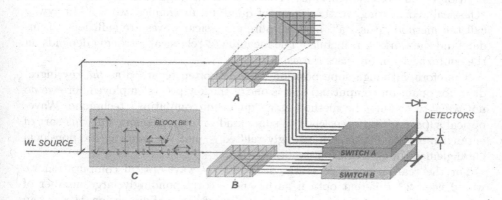

Fig. 7. Detection of constructive or destructive interference for bit 1 of one residue for the subtraction $C = A - B$. The N different path lengths are realized by using N diffraction gratings.

and 28 x 2 = 56 intensity detectors and 56 switches. Each switch is approximately a 100–to–1 switch. The disadvantage of the proposed method, therefore, is its complex way of implementation.

The speed of computation is — apart from the delay introduced by real switches and detectors for input and output of the result — completely given by the distance between switches and detector. Some kind of beam overlap (e.g. by a beamsplitter) has to be realised for a standard intensity detector between both elements. For interference based detection by four-wave mixing this is not necessary because we might introduce the lower path wave from behind. With integrated optics it might be anyway no problem to achieve distances below 50 μm which would correspond to 167 femtoseconds time of flight.

Of course, the high speed of the computation would only be usable if we would have a suitable overall computing architecture. Delivering the signals from and to the detectors and the switches in practice will introduce delays which will significantly reduce the achievable speed. Three–dimensional integrated optical systems, perhaps even based on photonic crystals, might make a future realization feasible. One might think of using such a technique within high-speed fiber networks, for example for self-routing of data packets[30] or high speed decryption. But at the moment, we do not see how a real implementation would lead to practical benefit.

Apart from that it is not clear what application really would need such high speed arithmetic. One has to keep in mind that for most high-speed numeric applications simple parallelization (done by real parallelization or by pipelining) is the best practical approach.

5   Classical, Wave-Optical, and Quantum Computing

The core idea to achieve the very fast computation is to generate the superposition of all possible solutions. This is what distinguishes the method from "classical" computing. In the jargon of quantum mechanics, we would anyway call the method "classical" because purely classical waves are sufficient to understand the working principle. (The meaning of "classical" strongly depends on the context within one uses the word.)

Therefore, although superposition (which is often regarded as *the* key ingredient for quantum computing) also is one of the key points employed here we do not want to classify the method as a "quantum computing" technique. Wave-optical superposition even leads to some kind of entanglement, but this sort of entanglement is — although strongly related — not the same as the non-local entanglement of quantum mechanics[31, 32].

For the realization of N (entangled) bits by wave-optical computation we would need 2^N different optical paths and a correspondingly large number of photons. As Londero et al. pointed out, a Hilbert space of dimension M x N can be realized by N particles with M states. For the wave-optical computation we only use one particle (one photon) that interferes with itself. But this particle can be in a superposition of M states (the different paths). Londero discusses that in general this leads to a scaling of the volume of the computing device and the number of detectors with $M \times N$. This seems to be true, but it does not imply that for most practical problems this upper bound is relevant.

Consequently, there are problems which in principle might be better solved by a quantum computer. But on the other hand, for other problems implementation of wave-optical computing might be much more simple and faster than achieving the same result by quantum computing. Both, wave-optical as well as quantum computing share the parallelization based on superposition. We want to emphasize that the wave-optical computing as described here is not used to somehow realize or emulate a quantum computing algorithm[33–37]. An interesting comparison of quantum and classical wave computing for Grover's database search algorithm[38] has been done by Lloyd[39, 40].

6 Conclusions

We have proposed a method for ultrafast digital-optical arithmetic. The minimum computation time is given by the time of flight of the distance between the input switches and the detectors. Contrary to other approaches, the switches can be located immediately in front of the detectors. Therefore, extremely high speed in the femtosecond region should be achievable. This high speed is only possible because we use the superposition of all possible computations and by coding the numbers by optical path lengths. The accuracy of the computations can be chosen freely by parallelization which is realized here by residue arithmetic.

At the moment we doubt much practical value because, first of all, the suitable application is missing, and, additionally, an overall computing architecture that could exploit the extremely high speed is not available. Anyway, the architecture might proof fruitful for optical computing devices in the future.

References

1. Tada, K., Arakawa, T., Noh, J., Haneji, N.: Ultra-wideband ultrafast multiplequantum- well optical modu(a)ltors. Photonics Based on Wavelength Integration and Manipulation IPAP Books 2, 199–212 (2005)
2. Wada, O.: Femtosecond all-optical devices for tera-bit/sec optical networks. Electron Devices Meeting, IEDM Technical Digest 589–592 (2000)
3. Taylor, H.: Guided wave electrooptical devices for logic and computation. Applied Optics 17(10), 1493–1498 (1978)
4. Psaltis, D., Casasent, D.: Optical residue arithmetic: a correlation approach. Applied Optics 18(2), 163–171 (1979)
5. Huang, A., Tsunoda, Y., Goodman, J., Ishihara, S.: Optical computation using residue arithmetic. Applied Optics 18(2), 149–162 (1979)
6. Bocker, R., Clayton, S., Bromley, K.: Elecrooptical matrix multiplication using the twos complement arithmetic for improved accuracy. Applied Optics 22(13), 2019–2021 (1983)
7. Mirsalehi, M., Shamir, J., Caulfield, H.: Residue arithmetic processing utilizing optical fredkin gate arrays. Applied Optics 26(18), 3940–3946 (1987)

8. Pahari, N., Das, D., Mukhopadhyay, S.: All-optical method for the addition of binary data by nonlinear materials. Applied Optics 43(33), 6147–6150 (2004)
9. Li, Y., Eichmann, G., Dorsinville, R., Alfano, R.: Demonstration of a picosecond optical-phase-conjugation-based residue-arithmetic computation. Optics Letters 13(2), 178–180 (1988)
10. Li, Y., Kim, D., Kostrzweski, A., Eichmann, G.: Conten-addressable-memorybased single-stage optical modified-signed-digit arithmetic. Optics Letters 14(22), 1254–1256 (1989)
11. Alam, M.: Parallel optical computing using recoded trinary signed-digit numbers. Applied Optics 33(20), 4392–4396 (1994)
12. Wu, W., Campbell, S., Yeh, P.: Implementation of a photorefractive arithmetic logic unit for multiwavelength information processing. J. Opt. Soc. Am. B 13(11), 2549–2557 (1996)
13. Datta, A., Munshi, S.: Signed-negabinary-arithmetic-based optical computing by use of a single liquid-crystal-display panel. Applied Optics 41(8), 1556 (2002)
14. Roy, J.N., Integrated, D.G.: All–optical logic and arithmetic operations with the help of a toad-based interferometer device - alternative approach. Applied Optics 46(22), 5304–5310 (2007)
15. McAulay, A.: Optical computer architectures for supervised learning systems. Computer 25(5), 72–75 (1992)
16. Sawchuk, A., Strand, T.: Digital optical computing. Proceedings of the IEEE 72(7), 758–779 (1984)
17. Sheppard, C.J.R., Roy, M., Sharma, M.D.: Image formation in low-coherence and confocal interference microscopes. Appl. Opt. 43, 1493–1502 (2004)
18. Oltean, M.: A light-based device for solving the hamiltonian path problem. In: Calude, C.S., Dinneen, M.J., Păun, G., Rozenberg, G., Stepney, S. (eds.) UC 2006. LNCS, vol. 4135, pp. 217–227. Springer, Heidelberg (2006)
19. Oltean, M.: Solving the hamiltonian path problem with a light-based computer. Natural Computing 7, 57–70 (2008)
20. Oltean, M., Muntean, O.: Solving the subset-sum problem with a light-based device. Natural Computing, 10.1007/s11047–007–9059–3 (2007)
21. Haist, T., Osten, W.: An optical solution for the traveling salesman problem. Optics Express 15, 10473–10482 (2007)
22. Haist, T., Osten, W.: An optical solution for the traveling salesman problem: erratum. Optics Express 15, 12627 (2007)
23. Dolev, S., Fitoussi, H.: The traveling beams optical solutions for bounded np-complete problems. In: Crescenzi, P., Prencipe, G., Pucci, G. (eds.) FUN 2007. LNCS, vol. 4475, pp. 120–134. Springer, Heidelberg (2007)
24. Jacques, V., Wu, E., Grosshans, F., Treussart, F., Grangier, P., Aspect, A., Roch, J.F.: Experimental realization of wheeler's delayed-choice gedanken experiment. Science 315, 966–968 (2007)
25. Swartzlander, E.: Digital optical arithmetic. Applied Optics 25(18), 3021–3032 (1986)
26. Mirsalehi, M., Shamir, J., Caulfield, H.: Three-dimensional optical fredkin gate arrays applied to residue arithmetic. Applied Optics 28(12), 2429–2438 (1989)
27. Heinrich, M., Athale, R., Haney, M.: Numerical optical computing in the residue number system with outer-produce lookup tables. Optics Letters 14(16), 847–849 (1989)
28. Cormen, T., Leiserson, C., Rivest, R., Stein, C.: Introduction to algorithms. MIT Press, Cambridge (2001)

29. Siez, S., Ludwig, R., Schmidt, C., Feiste, U., Weber, H.: 160 gb/s optical sampling by gain-transparent four-wave mixing in a semiconductor optical amplifier. IEEE Photonics Technology Letters 11(11), 1402–1404 (1999)
30. Cotter, D., Lucek, J., Shabeer, M., Smith, K., Rogers, D., Nesset, D., Gunning, P.: Self-routing of 100 gbit/s packets using 6 bit keyword address recognition. Electronics Letters 31(17), 1475–1476 (1995)
31. Spreeuw, R.: A classical analogy of entanglement. Foundations of Physics 28(3), 361–374 (1998)
32. Spreeuw, R.: Classical wave-optics analogy of quantum-information processing. Physical Review A 63, 062302-1–8 (2001)
33. Cerf, N., Adami, C., Kwiat, P.: Optical simulation of quantum logic. Physical Review A 57(3), R1477–R1480 (1998)
34. Kwiat, P., Mitchell, J., Schwindt, P., White, A.: Grover's search algorithm: an optical approach. Journal of Modern Optics 47(2/3), 257–266 (2000)
35. Bhattacharya, P.: Parallel optical image processing with image-logic algebra and a polynomial approach. Applied Optics 33(26), 6142–6145 (1994)
36. Londerao, P., Dorrer, C., Anderson, M., Wallentowitz, S., Banaszek, K., Walmsley, I.: Efficient optical implementation of the bernstein-vazirani algorithm. Physical Review A 69, 010302 (2004)
37. Kok, P., Murno, W., Nemoto, K., Ralph, T., Dowling, J., Milburn, G.: Linear optical quantum computing with photonic qubits. Review of Modern Physics 79, 135–174 (2007)
38. Grover, L.: Quantum mechanics helps in searching for a needle in a haystack. Physical Review Letters 79(2), 325–328 (1997)
39. Lloyd, S.: Quantum search without entanglement. Physical Review A 61, 010301-1–4 (1999)
40. Meyer, D.: Sophisticated quantum search without entanglement. Physical Review Letters 85(9), 2014–2017 (2000)

Photonic Reservoir Computing
with Coupled Semiconductor Optical Amplifiers

Kristof Vandoorne[1], Wouter Dierckx[1], Benjamin Schrauwen[2],
David Verstraeten[2], Peter Bienstman[1], Roel Baets[1],
and Jan Van Campenhout[2]

[1] Photonics Research Group, Dept. of Information Technology,
Ghent University – IMEC, Sint-Pietersnieuwstraat 41,
9000 Gent, Belgium
`kristof.vandoorne@UGent.be`
[2] PARIS, Dept. of Electronics and Information Systems,
Ghent University, Sint-Pietersnieuwstraat 41,
9000 Gent, Belgium

Abstract. We propose photonic reservoir computing as a new approach
to optical signal processing and it can be used to handle for example large
scale pattern recognition. Reservoir computing is a new learning method
from the field of machine learning. This has already led to impressive
results in software but integrated photonics with its large bandwidth and
fast nonlinear effects would be a high-performance hardware platform.
Therefore we developed a simulation model which employs a network of
coupled Semiconductor Optical Amplifiers (SOA) as a reservoir. We show
that this kind of photonic reservoir performs even better than classical
reservoirs on a benchmark classification task.

1   Introduction

Reservoir Computing was recently proposed [1,2] as a general framework to
handle classification and recognition problems. The reservoir itself consists of a
network of coupled nonlinear elements and their interactions facilitate the processing and classification of the incoming signals by the readout function. These
reservoirs were until now mainly software-based and they have been employed
successfully in a large variety of applications like speech recognition [3,4,5], event
detection [6], robot control [7], chaotic time series generation and prediction [1,8].

Photonic reservoir computing could be used for a large variety of problems,
going from large scale pattern recognition in (real-time) video data to signal
processing (header recognition, error correction, etc.) in optical fiber networks.
We studied a photonic reservoir, made of coupled SOAs, and this paper presents
the first results. The structure of the paper is as following. In section 2 we will
go deeper into this new concept of reservoir computing. The next section deals
with the (photonic) implementation aspects. Section 4 describes the classification task we used, to show the potential of photonic reservoir computing. This

S. Dolev, T. Haist, and M. Oltean (Eds.): OSC 2008, LNCS 5172, pp. 46–55, 2008.

task requires the reservoir to distinguish between a triangular and a rectangular waveform. It turns out that a photonic reservoir — with only a little bit of tuning and a limited number of 25 SOAs — can already distinguish between the two signals over 97 % of the time.

2   Reservoir Computing

2.1   Classical Approach

In this digital age, signals are often transferred to the digital domain for signal processing. Nature shows us however that there are alternatives, which can be superior for complex classification and recognition problems, like the human brain combined with eyesight. Machine learning looks to the biological world for inspiration, where organisms often learn from their failures and successes or in other words from examples. Systems in machine learning are accordingly trained to perform certain tasks. Artificial Neural Networks (ANN) are an example of such a system and take the analogy with the biological world one step further [9]. The inspiration for the system comes from its biological counterpart, the human brain, which consists of neurons. The human brain lacks speed compared to a computer, but it compensates this by having a rich interconnection topology. Each connection has a certain weight attached to it and these weights can be adapted during the training process.

Feed-forward neural networks have been extensively used for non-temporal problems and they are well understood due to their non-dynamic nature. At the same time, this limits their applicability in dealing with time varying signals. Indeed, neural networks with feedback loops (so-called recurrent neural networks) provide some kind of internal memory which allows them to extract time correlations. However, this turned out to be a hurdle in finding a general learning rule, which is a method used to train the neural network to perform a desired task. This is why different rules exist for different tasks and topologies, thus limiting their broad applicability.

Around 2002 two solutions (Liquid State Machines and Echo State Networks) were independently proposed [1,2]. They have in common that the network is split up in two parts. One part — the reservoir — is a random recurrent neural network that is left untrained and kept fixed while using. The input is fed into this reservoir. The second part is a readout function, which takes as input the reservoir state — a collection of the states of the individual elements. In order to be able to achieve useful functionality, this part of the system needs to be trained, typically on a set of inputs with known classifications. This process is visualized in figure 1. Any kind of classifier or regression function could be used, but it turns out that for most applications a simple linear discriminant suffices. In this way the interesting properties of recurrent neural networks are kept in the reservoir part, while the training is now restricted to the memoryless readout function.

One might wonder why such an approach would be useful to solve complex classification tasks. However, it is well known in the machine learning community that projecting a low-dimensional input into a high-dimensional space can actually be beneficial for the performance of a classification algorithm. As classes, which are only separable by a nonlinear function in the low-dimensional space, can become separable by an easier, linear function in the high-dimensional space [10]. This concept is applied e.g. in support vector machines.

The reservoir could be seen as an integration of the temporal correlations in the signal into a spatial correlation in the reservoir state. This is not to say that any recurrent neural network will do. Rather, it appears that the dynamics of the network should be in the dynamic region which corresponds to the edge of stability [11]. The dynamics depend on the amount of gain and losses in the network and they should be balanced. If the network is over-damped there is no memory inside the reservoir, if it is under-damped the network will react chaotically.

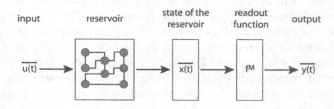

Fig. 1. Reservoir Computing

Recently a toolbox, able to simulate and test reservoirs, was created [12]. In this toolbox the reservoirs are neural networks. One of those is the classical variant where the signals are analog and every node is a hyperbolic tangent function operating on a weighed sum of its inputs. This function is S-shaped as in figure 2 (left). In this kind of network the nodes themselves are very simple, while the dynamics come from the complex interconnection topology.

2.2   Photonic Approach

The theory behind reservoir computing is not restricted to neural networks. One requirement for the reservoir is fading memory, which means that the influence of an input should fade away slowly. The present software implementations are rather slow and therefore we investigated the potential of a hardware implementation based on light. This could be faster and more power efficient due to the large bandwidth and fast nonlinear effects inherent to light.

Due to the nature of reservoir computing, its implementation can be split up in two distinctive parts: the reservoir on the one hand and the readout function on the other. Since the computational power of reservoir computing seems to reside mainly in the reservoir due to its feedback and nonlinearities, the focus of the research was on a photonic reservoir. As mentioned before the readout is a simple linear discriminant, but its training depends on mathematical calculations like matrix inversion. This could initially be done off-line with a computer or by an electronic chip.

Because the specifications for the reservoir do not seem to be very rigid, the choice of possible nanophotonic components was vast. We opted for coupled Semiconductor Optical Amplifiers (SOA), based on two observations. First of all, the steady-state curve of an SOA resembles the S-curve used in analog neural networks (figure 2) — at least for the upper branch, but since optical power is non-negative this is a restriction we have to cope with. This resemblance made SOAs more likely to be able to bridge the reservoir and the photonic world. The dynamic behavior of an SOA is however more complex in comparison to the classical analog implementations. The carrier dynamics come into play at high data rates and because of this a reservoir of SOAs is a middle ground between simple nodes with a complex network (the classical tanh reservoirs) and one very complex node. Second, SOAs are broadband which makes the communication between different nodes less critical as would be the case with resonating structures.

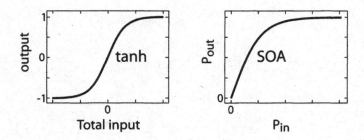

Fig. 2. (Left) tanh for analog ANN — (right) SOA: steady state

3   Simulation Model

We developed our simulation program for photonic reservoirs within the framework of the toolbox, mentioned previously. This allows us to use the existing training schemes for the memoryless readout function. For further details about this open source toolbox we refer to the manual online [12] and the article by D. Verstraeten et al. [13].

3.1   SOA Model

In our simulations we work with a traveling wave SOA. This kind of SOA has anti-reflection coatings on its facets, which allows us to neglect the influence of reflections. We use the standard traveling wave SOA equations [14]. We neglect the influence of Amplified Spontaneous Emission (ASE) and spectral hole burning in this model. This means that we assume that the input signal itself will be strong enough to dominate the ASE and that we only use light at one wavelength. To incorporate the longitudinal dependence of the gain, the equations can be solved for a concatenation of small sections of the SOA. Since the latter can be time consuming when working with large networks of SOAs, we work mainly with one section. Moreover since reflections are neglected at this stage, we use unidirectional signal injection. This reduces the number of rate equations to be solved to one.

3.2   Topology and Reservoir Simulation

The classical reservoir implementations with neural networks have random interconnection topologies. Since the standard optical chip is still 2D we investigated structures that can be realized without intersections. Two of those structures are depicted in figure 3. The left structure is a waterfall system which acts as a nonlinear delay-line. Although this feed-forward topology is relatively simple, it has already been successfully used to model nonlinear systems [15]. The other network has feedback connections on the sides in order to avoid crossings. Since the SOAs are modeled as unidirectional, the connections are too.

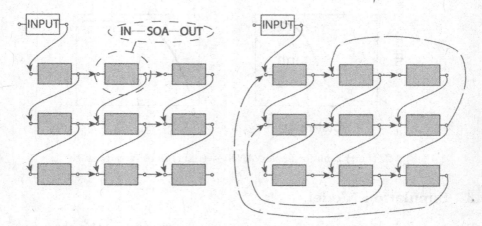

Fig. 3. Two topologies: (left) a feed-forward network – (right) a waterfall network with feedback connections (long dash) at the edges

At every time instant two computational steps are taken. During the first step the internal state of every node is updated, while during the second step the outputs are transferred to the inputs they are connected to. The splitters and combiners are modeled as adiabatic and every connection can have a different delay and attenuation.

The readout function takes as input the power of every SOA in the network at every time instant. This is the basic structure of the simulation model. Next, we will look at tasks that can be solved with these kind of networks.

4   Pattern Recognition

4.1   Task Description

We will use a simple classification task to demonstrate the potential of photonic reservoir computing. The task is depicted in figure 4. The system has to be able, by means of training by examples, to instantly differentiate between a rectangular and a triangular waveform. Moreover, if the input signal changes the system has to change its output as fast as possible. In the top part (a) of figure 4, an example of such an input is depicted. Figure 4b shows the output that the system should generate accordingly. If the input is triangular than the system should constantly return 1, if the input is rectangular it should return -1. Figure 4c shows the state of a few SOAs as they are excited by the input, while figure 4d shows the result of the readout function. The readout uses a linear combination of the states of the reservoir nodes, to approximate the desired output (black curve) as closely as possible (blue curve). In the last stage a sign function is used on this approximation to define the final output of the system. As a result the output is either 1 or -1 as can be seen in figure 4e. In the example the system manages most of the time to follow the desired output.

Since the output function is memoryless, it should be able to handle transitions of the waveform at different instants. Hence several samples are made with different transitions at different instants. One part of these samples is used to train the readout function, while the other part is used to test it. These test results are used to define the quality of the reservoir.

4.2   Results

In figure 5 some results are displayed from our simulations. The vertical lines are error bars, which show the standard deviation on the reservoir performance over ten runs. The variation comes, for the photonic reservoirs, from different samples with different transitions at different instants. The tanh networks have an extra variation source because they are randomly created.

In the left figure the two photonic topologies from figure 3, with 25 SOAs, are compared against the attenuation in the connections. This attenuation influences the dynamic regime and the higher the attenuation, the more damped the system is. It shows that feedback is beneficial for the performance of photonic reservoirs, when used in the appropriate dynamic regime. The best result is an error rate of 2.5 % for the network with feedback loops. This means that the reservoir is only in 2.5 % of the time incorrect in its distinction between the rectangular and the triangular waveform. If the attenuation in the connections gets too small then the performance decreases dramatically for the feedback network.

On the right of figure 5 a feedback reservoir with SOAs and the classical tanh reservoirs are compared, each with 25 nodes. Since there is feedback in both networks we use the spectral radius ($\rho(\cdot)$) as a measure for the dynamics in the system. The spectral radius is the absolute value of the largest eigenvalue of the connection matrix C, containing all the gain and loss in the network and is an

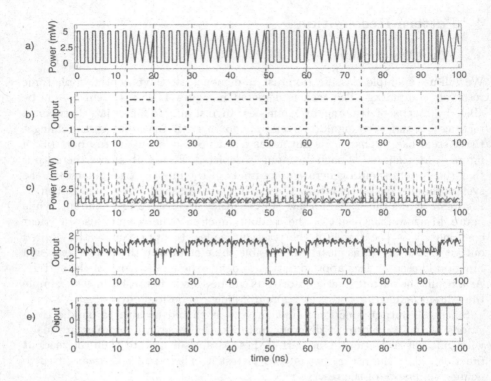

Fig. 4. Pattern recognition task: a) Input signal with different transitions between the rectangular and triangular waveform b) desired output c) state (i.e. optical power level) of some of the reservoir nodes d) The approximation (blue) of the desired output (black) by the readout function, e) final output of the system (red)

often used parameter in the field of reservoir computing. In a linear network a spectral radius smaller than one means the network is stable, a value larger than one means chaotic. The interesting dynamical region, the edge of stability, holds for spectral radii just below one. Although our network is nonlinear, we can still use this as an approximation, where the spectral radius acts as an upper estimate. The gain for every node is linearized around zero input power and for a connection matrix C with eigenvalues $\lambda_i, \ldots, \lambda_n$ this leads to the following spectral radius calculation:

$$\rho(C_{lin(0)}) = \max_{1 \leq i \leq n} |\lambda_i| \tag{1}$$

The classical reservoir appears to behave better for small spectral radii (when the system is damped) with an optimum value around 3.5 %. As soon as the spectral radius gets too high, the system becomes chaotic which explains the large error bars and bad results. The same holds for the photonic reservoirs at higher spectral radii. The curve for the SOA network with feedback is the same in the two figures but plotted against a different parameter. An optimal value of

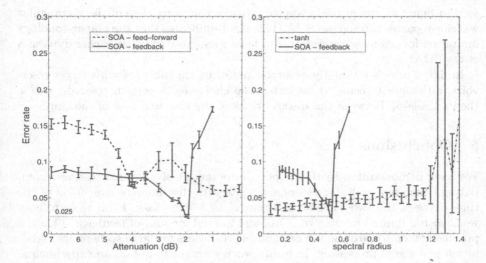

Fig. 5. Results with frequency of 0.5 GHz, simulation time: 100 ns, amplitude power: 5 mW, delay: 6.25 ps, fixed input–output weights, SOA input current: 80 mA, 25 nodes: (left) – photonic reservoirs with and without feedback, (right) – classical reservoirs versus photonic reservoirs with feedback

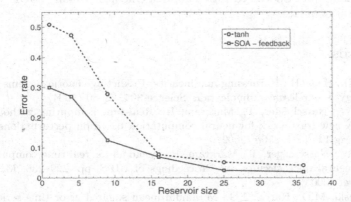

Fig. 6. This figure shows the influence of the reservoir size on the performance of the classical and photonic reservoirs (with feedback)

2.5 % is obtained for a spectral radius around 0.5. This means that the system detects the correct waveform almost 97.5 % of the time.

It is remarkable that the photonic reservoir with feedback is slightly better than the classical reservoir for this task, considering the simple photonic topology. One explanation is the different response of an SOA to different rise times. The rectangular waveform, although smoothed with a tanh, rises faster than the triangular waveform, causing a depletion of the carriers in the SOA. This can be seen in figure 4c, when peaks appear whenever a rising edge of the rectangular waveform passes through an SOA. This result indicates that the planar-topology limitation for photonic reservoirs is at least compensated by the richer dynamics of the SOAs.

In figure 6 we see that the reservoir performance enhances with larger reservoirs, although it seems to saturate. The choice for a certain reservoir size is then a trade-off between the specifications of the task and cost of the chip.

5   Conclusions

We have demonstrated in this paper the potential of photonic reservoir computing, since our photonic reservoirs manage to discriminate over 97 % of the time between the two waveforms in our classification task. Even though, they work with a limited number of SOAs and limited amount of feedback. This is a promising step toward the use of photonic reservoirs for large scale image recognition and signal processing. In future work we want to obtain an experimental verification of the described simulation results.

Acknowledgments

KV acknowledges the Special Research Fund of Ghent University (BOF – UGent) for a doctoral mandate.

References

1. Jaeger, H., Haas, H.: Harnessing nonlinearity: Predicting chaotic systems and saving energy in wireless communication. Science 304, 78–80 (2004)
2. Maass, W., Natschläger, T., Markram, H.: Real-time computing without stable states: A new framework for neural computation based on perturbations. Neural Computing 14, 2531–2560 (2002)
3. Maass, W., Natschläger, T., Markram, H.: A model for real-time computation in generic neural microcircuits. In: Proceedings of NIPS, pp. 229–236. MIT Press, Vancouver (2003)
4. Skowronski, M.D., Harris, J.G.: Minimum mean squared error time series classification using an echo state network prediction model. In: Proceedings of IEEE International symposium on circuits and systems. Institute of Electrical and Electronics Engineers, Island of Kos (2006)
5. Verstraeten, D., Schrauwen, B., Stroobandt, D., Van Campenhout, J.: Isolated word recognition with the Liquid State Machine: a case study. Information Processing Lett. 95, 521–528 (2005)
6. Jaeger, H.: Reservoir riddles: Suggestions for echo state network research (extended abstract). In: Proceedings of IEEE International Joint Conference on Neural Networks, pp. 1460–1462. Institute of Electrical and Electronics Engineers, Montreal (2005)
7. Joshi, P., Maass, W.: Movement generation and control with generic neural micrrocircuits. In: Proceedings of BIO-ADIT, pp. 16–31. Springer, Berlin (2004)

8. Steil, J.J.: Online stability of backpropagation-decorrelation recurrent learning. Neurocomputing 69, 642–650 (2006)
9. Bishop, C.M.: Neural Networks for Pattern Recognition. Clarendon Press, Oxford (1995)
10. Vapnik, V.N.: An overview of statistical learning theory. IEEE Trans. Neural Networks 10, 988–999 (1999)
11. Legenstein, R., Maass, W.: What makes a dynamical system computationally powerful? In: New directions in statistical signal processing: from systems to brain, pp. 127–154. MIT Press, Cambridge (2007)
12. Reservoir lab: Reservoir Computing Toolbox, http://snn.elis.ugent.be/node/59
13. Verstraeten, D., Schrauwen, B., D'Haene, M., Stroobandt, D.: An experimental unification of reservoir computing methods. Neural Networks 20, 391–403 (2007)
14. Agrawal, G.P., Olsson, N.A.: Self-Phase Modulation and Spectral Broadening of Optical Pulses in Semiconductor-Laser Amplifiers. IEEE J. Quantum Electron. 25, 2297–2306 (1989)
15. Cernansky, M., Makula, M.: Feed-forward echo state networks. In: Proceedings of IEEE International Joint Conference on Neural Networks, pp. 1479–1482. Institute of Electrical and Electronics Engineers, Montreal (2005)

Electro-Optical DSP of Tera Operations
per Second and Beyond
(Extended Abstract)

Dan E. Tamir[1], Natan T. Shaked[2], Peter J. Wilson[3], and Shlomi Dolev[4]

[1] Department of Computer Science, Texas State University, San Marcos,
Texas 78666, USA
dt19@txstate.edu
[2] Department of Electrical and Computer Engineering, Ben-Gurion University of the Negev,
P.O. Box 653, Beer-Sheva 84105, Israel
natis@ee.bgu.ac.il
[3] Kiva Design, 104 Forest Trail, Leander, Texas 78641, USA
pete@kivadesigngroup.com
[4] Department of Computer Science, Ben-Gurion University of the Negev, P.O. Box 653,
Beer-Sheva 84105, Israel
dolev@cs.bgu.ac.il

Abstract. We present a vector-by-matrix multiplier architecture incorporating the high potential of optical computing within electronics. This architecture stems from advances in optical switching technology, optical communication, and laser on silicon, and overcomes previous bottlenecks implied by the speed of transferring information to the optical vector-by-matrix multiplier. Based on this architecture, we present in detail a feasible electro-optical DSP co-processor that can obtain more than 16-Tera integer operations per second. The use of the new architecture for several principal DSP applications is detailed, showing a significant improvement of at least two orders of magnitudes, over existing DSP technologies. Examples of possible applications, including motion estimation engine, string matching, and geometry engine systems, are provided. In addition, the architecture enables an improvement of previous solutions to bounded NP-complete problems by extensively reducing the size of the solver, while preserving an efficient computation time.

Keywords: Optical Computing, Parallel Processing, Digital Signal Processing, Traveling Sales Person Problem, Bounded NP-Complete Algorithms.

1 Introduction

The need to double the computation speed every *18* months (Moore law) or less still exists. To cope with the frequency limitations of VLSI technologies and still satisfy this need, the microprocessor industry proposes multi-core technologies, essentially suggesting the use of parallel computations. New parallel processing paradigms are being considered and evaluated due to the need for fast communication capabilities among the cores. The revolution in microprocessor architectures can open new opportunities for the use of optical computing. Several new architectures were recently suggested [1-5].

S. Dolev, T. Haist, and M. Oltean (Eds.): OSC 2008, LNCS 5172, pp. 56–69, 2008.

These architectures are based on optical communication over silicon and utilize the fact that no cross-talks occur in optics, just like the case of the widely-used fiber optics communication.

In this work, we present a new electro-optical architecture that enables a fast vector-by-matrix multiplication. In addition, we present new solutions that enhance the capabilities of this architecture as a co-processor dedicated for DSP-intensive and computationally-intensive applications. The fact that optical matrix multiplication can be carried out in parallel and in a very fast rate must be supported by electronic interface for transferring data to the multiplier in a way that utilizes the multiplier capabilities. Otherwise, the speed of the computation is determined by the data transfer bottleneck.

The operation of a vector-by-matrix multiplication is involved in several computationally-intensive applications such as rendering computer-generated images, beam forming, radar detection, and wireless communication systems. Many of these practical applications require efficient and fast implementation of vector-by-matrix multiplication.

In addition, vector-by-matrix multiplication can be used as a building block for numerous DSP procedures such as convolution, correlation, and certain transformations, as well as distance and similarity measurements [6,7,8]. For example, a discrete Fourier transform (DFT) can be implemented as a special case of vector-by-matrix multiplication.

The electro-optical vector-by-matrix multiplier (VMM) presented in this paper can perform a general vector ($1\times256\times8$ bit) by matrix ($256\times256\times8$ bit) multiplication in one cycle of 8 nanoseconds (that is, at a rate of 125 MHz). This rate is faster than all other VMMs available today and it is expected to improve with the introduction of new electro-optical technologies. In addition, the proposed VMM can serve as a co-processor attached to a DSP or a RISC CPU (referred to as a controller) and significantly enhance the performance of the controller.

Previous commercial electro-optical VMMs (e.g. [1]) were limited to the rate of the electrical driver of the spatial light modulator (SLM) used to represent the VMM matrix. The present paper proposes an alterative architecture where the SLM and its electrical driver are not limited to low rates. Hence, the proposed VMM utilizes the optical advantages more efficiently.

Section 2 introduces the proposed electro-optical design. Section 3 presents possible implementation and provides data that show the design feasibility, as well as explains how this design can be improved along with technology advances. Section 4 presents and analyzes several applications, and Section 5 concludes the current paper.

2 The VMM Electro-Optical Unit

The ability to perform mathematical operations in free space (in the open space, with no wiring), in parallel, and without mutual interactions among the various signals is only a part of the inherent features of optical data processing. These features are utilized in this paper, as well as in many other optical VMM configurations that have been suggested in the technical literature [9-12]. The proposed VMM has two main components: the optical unit and the electrical driver. The next two sections elaborate on each of these two components respectively.

2.1 The VMM Optical Unit

Several configurations can be utilized to implement the optical component of the VMM. One of these configurations is based on the Stanford multiplier principle [12] illustrated in Fig. 1. As shown in this figure, the input vector of the VMM is represented by a set of light sources, the matrix of the VMM is represented by a slide mask or a real-time SLM and the output (multiplication-product) vector of the VMM is represented by a set of sensitive detectors. The light from each of the sources is spread vertically so that it illuminates a single column in the matrix and then each row in the matrix is summed onto a single detector in the detector array. This VMM configuration can be performed by several optical techniques. One of these techniques uses two sets of lenses. Each of these sets contains a cylindrical lens and a spherical lens and each of these lenses has a focal length of f. A single set of lenses has an equivalent focal length of $f/2$ in vertical/horizontal direction and f in the other direction. As shown in Fig. 1, the first set of lenses is positioned between the input vector (represented by the light sources) and the matrix (represented by the SLM). This set of lenses is positioned so that the light coming from each of the sources illuminates only a single column in the matrix, which means collimating the light diverging vertically from each of the sources, but imaging it in the horizontal direction.

2.2 The VMM Electrical Driver

Figures 2 and 3 present high level descriptions of the electrical components of the system. A possible implementation of the system outlined in these figures is presented in Section 3.

Figure 2 shows the VMM electrical driver. The driver is comprised of 256 single electrical driver (SED) units and has two types of inputs: a $1 \times 256 \times 8$ bit input vector A which is the VMM input vector (which is the VCSEL source array driving signal), and a set of 256 vector inputs B_0 to B_{255} (total of $256 \times 256 \times 8$ bit). The output of the VMM is a $1 \times 256 \times 20$ bit vector C. The VMM output vector C is an aggregation of the set of scalar outputs (C_0 to C_{255}), where the output C_j emerging from SED_j is a single 20-bit bus which is yielded by the output detector array.

Figure 3 shows the design of one of the $1 \times 256 \times 8$ bit SED units. The input vector B_j can be stored in an internal dual-port modular memory buffer / shifter before being directed to the SLM. The output C_j is directed to the external output.

The configuration depicted in Fig. 3 supports a dot-product operation between the input row vector A (being converted into light by the VCSELs) and one column of the entire SLM matrix. Each SED unit performs one vector dot-product operation per cycle of 8 ns (the reciprocal of 125 MHz). Combined together, the 256 units perform a vector ($1 \times 256 \times 8$ bit) by matrix ($256 \times 256 \times 8$ bit) multiplication operation at a rate of 125 MH.

Each of the units depicted in Fig. 3 is capable of executing 256×125 million multiply-accumulate instructions per second. This is equivalent to 64 Giga integer

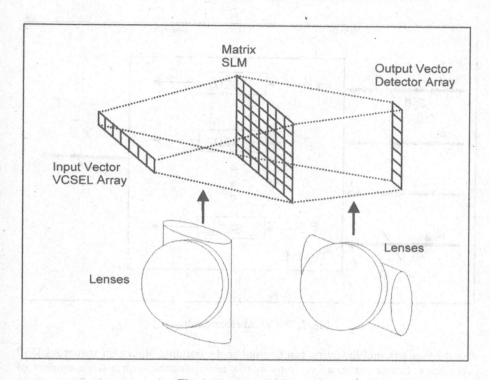

Fig. 1. The Stanford VMM

operations per second (GIPS). Since the entire VMM system consists of *256* units that are the same as the one unit depicted in Fig. 3, it can perform in peak performance of *16384* GIPS.

The proposed VMM can serve as a co-processor attached to a DSP or a RISC CPU, referred to as the controller. In addition, the controller can utilize other co-processors. The entire system (controller, co-processors and VMM) is depicted in Fig. 4. The figure shows an example where one of the additional co-processors is capable of shuffling vector elements. This may be useful for implementing convolution and correlation. It is assumed that in the typical mode of operation, the controller, along with other co-processors, prepares an input vector and an input matrix to be sent to the VMM for processing. We refer to this as pre-processing. The output of the VMM is sent back to the controller, where it may go through additional processing (referred to as post-processing) before being sent to other devices or back to the VMM. Sound pre-processing and post-processing operations can reduce the amount of VMM operations. For example, reusing the same input vector as much as possible, while altering only some of the matrix values, can reduce the amount of communication between the controller and the VMM and enable the VMM to operate at a peak rate that is greater than *125* MHz.

The VMM is capable of performing one generic operation. That is, a vector (*1×256×8* bit) by matrix (*256×256×8*) multiplication. Nevertheless, special cases of

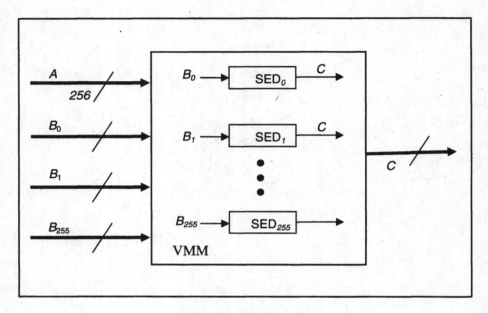

Fig. 2. The VMM electrical driver

vector-by-matrix multiplication can be used as the building blocks for numerous DSP procedures. For example, a vector-by-matrix multiplication with a large number of elements (more than 256 or 256×256 elements), vector-by-matrix multiplication with a small number of elements (less than 256 or 256×256 elements), various matrix-by-matrix multiplications, dot-product operations including convolution, correlation, transforms (e.g. discrete Fourier / cosine transforms), and L2 norm operations, as well as addition, complex, and extended-precision operations. These operations, which are the building blocks of numerous DSP-intensive applications, are further analyzed in a technical report [13].

3 Implementation of the VMM Electrical Driver

Figures 3, 5, and 6 illustrate the components of the VMM electrical driver. These figures include several implementation details showing that it is feasible to construct the system using existing mass-production VLSI technology.

As mentioned above, the driver consists of 256 SED units of 1×256×8 bit each (one of which is shown in Fig. 3). Several SED units are integrated into ALU chips, which are placed on an interface board. A set of SED units integrated into one chip is referred to as the ALU chip (see Fig. 5). The interface board contains several ALU chips (see Fig. 6). Overall, the interface board contains 256 SED units. There is design flexibility with respect to the number of SED units per ALU chip. This number dictates the required number of ALU chips in the interface board. After analyzing current technology, we have decided to investigate a configuration with 16 ALU chips, each of which contains 16 SED devices. This configuration is further analyzed in the next subsections.

3.1 The SED Unit

Each SED unit contains a relatively small dual-port modular memory of several thousands bytes (say *2048* bytes) and a *256* element SLM (i.e., *256* SLM transistors).

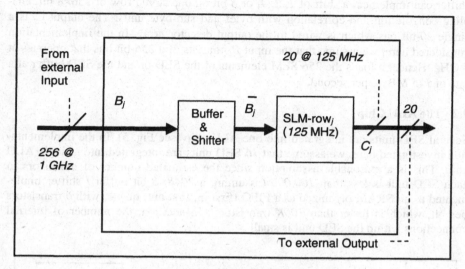

Fig. 3. A *1×256* SLM electrical driver

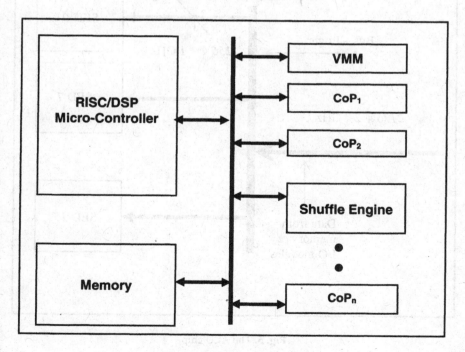

Fig. 4. System architecture

The SLM element of SED$_j$ (depicted in Fig. 3) represents row j of the SLM matrix. The input vector B_j can be stored in the internal buffer / shifter before being directed to the SLM. The buffer drives the SLM either directly with the input B_j or with a set of 256 bytes that has been previously stored in the buffer (B_j). In this case, the shifter can implement a shift of 1, 2, 4, or 8 bit on one stored row of 256×8 bit, enabling convolution and correlation with bytes and sub-byte units. The output C_j is a single 20-bit bus which is wired to the output detector array. In the implementation considered here, we assume that the input B_j consists of a 256-bit bus that operates at 1 GHz. Hence, it loads the 256 SLM elements of the SED or/and the SED buffer at a rate of 125 MByte per second.

3.2 The ALU Chip

Several SED units are integrated into one ALU chip (see Fig. 5). In the implementation investigated here, we assume that 16 SED units are integrated into a single ALU chip. This is a reasonable assumption since the estimated number of transistors in each SED unit is less than 100,000 (assuming a 2048×8 bit buffer / shifter implemented as an SRAM organized as a FIFO (first in, first out) queue; with 6 transistors per bit, which is fewer than 100K transistors). In addition, the number of internal connections within the SED unit is small.

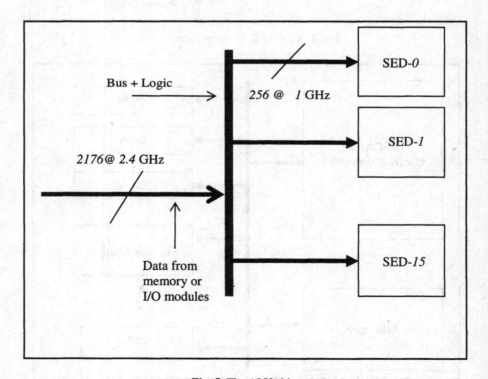

Fig. 5. The ALU chip

An ALU chip is connected to external memory or I/O devices through an external bus. An internal bus distributes the data to the 16 SED units inside the ALU chip. Each unit gets *256* bit at a rate of *1* GHz, hence a rate of *125* MByte per second. This means that the entire SLM matrix is updated at the same rate as the input row vector (*125* MHz).

Theoretically, a bus of *2048* bit operating at *2* GHZ is needed in order to supply *256* bit of data to each SED unit and sustain a rate of *125* MHz. Practically, we can use an internal bus of *2176* bit, which operates at *2.4* GHz, to drive the individual SED units. The reason for using a bus of *2176* bit is due to the need to synchronize the input data. It is assumed that each set of *16* data lines are synchronized through one synch line. Hence, *2048* data lines require *128* synch lines and the total number of input lines is *2176*. In addition, it is assumed that *20%* redundancy in the number of input bits is required in order to supply error correction and handshaking mechanism. This can be achieved by raising the frequency of the bus to *2.4* GHz. While this is a relatively wide bus operating at a relatively high frequency, it is well supported by current *0.65*-nm CMOS technology, where chips such as the IBM Cell contain thousands of pins, thousands of bus lines, and operate at rates that are above *3* GHz [14]. It is also in line with the ITRS 2005 / 2006 reports [15]. Furthermore, this bus is not a true multi-drop bus; rather, it is implemented as a number of narrower point-to-point interconnects (driving a very wide bus quickly and supporting many sinks would be a design challenge).

The proposed device requires a package with several thousand input pins. This is within the state of the art. As predicted in Ref. [16], packages with over *3000* connections are currently available.

Within the devices, several sources distribute Gigahertz-rate signals to several destinations. This is also within the capabilities of current technology. As an example, the Tilera TILE 64 implements *64* processors in a *90*-nm CMOS technology [17]. Each processor connects to five point-to-point networks implemented as unidirectional *32*-bit interconnects, which provides an aggregate bandwidth of *1.28* Terabit per second. This bandwidth is provided by *320* bits (lines). Hence, each wire in this older CMOS technology is providing *1280/320=4* Gigabit per second. Furthermore, Intel has described a *5* GHz on-chip network [18].

3.3 The Interface Board

The interface board depicted in Fig. 6 contains *16* ALU chips. In addition, the interface board contains lines that enable transfer of external inputs to the ALU chip. Each ALU chip is fed with data independently of other ALU chips. While there are no internal connections or dependencies between the ALU chips, they work in a source-synchronous mode, where each ALU chip operates at a rate of *2.4* GHz.

The interface board is comparable in complexity to a small-size InfiniBand interconnect board [19]. In fact, in some applications, it is conceivable that the ALU chips occupy more than one interface board. In these cases, the system is equivalent in complexity to several system-on-chip (SOC) units connected to an InfiniBand board. In addition, since there are no dependencies / communication transactions between ALU chips, the interface board does not require a fully-connected switch.

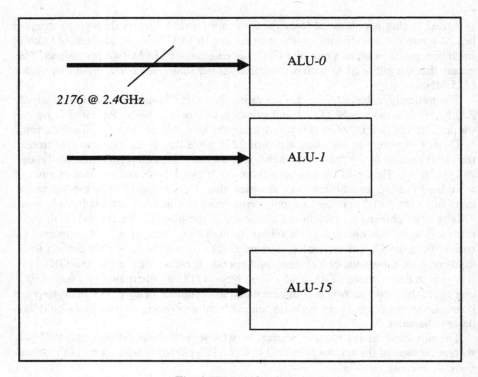

Fig. 6. The interface board

The conclusion from the above discussion is that the level of complexity of the interface board is moderate and that this design can be implemented with current of-the-
shelve VLSI technology.

3.4 System Synchronization

We consider two synchronization tasks. First, the SED units within an ALU chip have
to be synchronized. Second, the ALU chips within an interface board have to be
synchronized.

Synchronization of SED units within the ALU chip is relatively straightforward.
The SED units operate at *1* GHZ. Each SED contains small amount of logic and the
data path within the SED is short (*4* transistors in serial, one of which is an SLM
transistor). This is well below the characteristics of current multi-core systems which
contain *16* to *32* cores, a data-path with *11* (or more) transistors, and operate at frequencies of *3* GHz and above [17, 18, 20, 21].

Given the characteristics of an ALU chip, synchronizing the ALU chips within the
interface board–requires a relatively modest level of complexity. The synchronization
can be done by distributing a clock signal through the ALU chips. Commercial systems
that distribute clock through a clock-tree to hundreds of devices with skew of less than
50 picoseconds (ps) are available. Hence, a 2-branch *8*-level clock distribution unit can
be used to synchronize the *16* ALU chips. Using this tree, an uncertainty of the order of

less than 50 ps is achievable [17]. Nevertheless, since the source array operates at a rate of *125* MHz, *50* ps is a negligible uncertainty and can be factored into the design. The effect of uncertainty can cause a loss of one least significant bit from an entire set of *256×8* bit. Given the fact that we can use guard bits in the VMM, this uncertainty is tolerable.

4   DSP-Intensive and Computationally-Intensive Applications

The DSP procedures, listed in Section 2 and detailed in a technical report [13], can be used as building blocks for several DSP-intensive applications. As explained in the abovementioned technical report, the proposed architecture is capable of performing *125* million operations - of a (*1x256x8* bit) vector by a (*256x256x8* bit) matrix multiplication - per second. It can complete *32* billion cross correlations - of a (*1x256x8* bit) vector by a (*1x256x8* bit) vector - per second. In addition, the VMM can complete *125* million convolutions - of up to *511* samples with a finite input response filter of up to *256* taps - per second, and *31.25* million discrete Fourier transforms - of *256* complex samples - per second.

Goren *et al* [1] have elaborated on wireless-related processing, such as rake receiver and multi-user detection. Their proposed VMM, however, is at least two orders of magnitude slower than the current proposed VMM. In this section, we analyze the performance of the proposed system as a component in several other DSP-intensive and computationally-extensive applications.

4.1   The VMM as a Motion Estimation Engine

The VMM can be used to implement MPEG/H.264-like motion estimation (ME). Consider a current-frame macro-block of *16×16* pixels stored in the input vector and a search window of *32×48* pixels. The controller loads the SLM matrix with consecutive macro-blocks from the search window and the VMM implements cross-correlation with *256* blocks in each cycle (32 billion cross correlations per second). The entire *32×48* window contains *1536* instances of overlapping macro-block. Hence, it can be loaded into the SLM in *6* cycles of operation or at a rate of $125/6 = 20.8\overline{3}$ MHz. This is the rate at which the VMM can complete the search. The controller has to find the maximum of the cross-correlation function calculated by the VMM. Some variants of the above analysis can be of interest. For example, if an "informed" search, such as multi-resolution (pyramid) search, is implemented, then each stage can use *256* macro-blocks from the search window. A *3*-stage search in a *32×48* window at *1/4*-pixel resolution can be accomplished at *25* MHz per macro-block. A 1/8-pixel resolution requires one more cycle and can be accomplished in $125/6 = 20.8\overline{3}$ MHz [13].

4.2   String Matching Using the VMM

The motion estimation method described above is a special case of two-dimensional string matching. The VMM has exceptional capability to support string matching. It can be used for exhaustive string matching or to support advanced matching techniques such as the Boyer-Moore, Knuth-Morris-Pratt, or Rabin-Karp algorithms [22]. The string or substring to be matched is stored in the VMM input vector. It can be matched via cross-correlation with *256* other substrings stored in the SLM matrix. Under the current architecture, the VMM can sustain "row" (exhaustive) string matching rate of *256* Gigabit per second. This is at least two orders of magnitude faster than existing hardware architectures of exhaustive search. In the future, we plan to investigate the possibility of loading the SLM matrix at a rate that is higher than *125* MHZ. This can increase the row-string-matching capability of the VMM to the order of Terabit per second. In addition, we plan to investigate the VMM capability to support Basic Local Alignment Search Tools (BLAST) and algorithms for matching nucleotide or protein sequences [23].

4.3 The VMM in a Geometry Engine System

The VMM can be used to support the geometry pipeline of a computer-graphics system, where multitudes of polygons (generally triangles), represented by vertices in a *4*-dimensional homogeneous coordinate space, are subject to affine transformation and perspective/parallel projections. Affine transformations and perspective projections of the polygons require multiplication of *1×4* vectors, representing polygon vertices, by *4×4* matrices. The proposed VMM can complete *2* billion multiplications of a *1×4×8*-bit vector by a *4×4×8*-bit matrix per second. Hence, it can compute the transformation of *666* million triangles per second. The vertices, however, are represented by low resolution *8*-bit elements. To support *32*-bit floating point operations, i.e., operations with *24*-bit mantissa, the VMM has to enable *24*-bit multiplication. This can be achieved by generating *9* partial products. Reuse of data stored in the matrix can enable a rate of *222* million triangles per second. In this case, however, the controller is expected to do extensive pre-processing and post-processing (e.g., shift-and-add of partial products). This may require another dedicated co-processor. Furthermore, each *8*-bit×*8*-bit multiplication must produce no errors, since these errors can appear in significant bits of the *24*-bit result. Thus, the VMM should be capable of generating an exact *16* bit result. This can be accomplished by adding guard bits to the VMM multiplication unit.

4.4 Bounded NP-Complete Problem Optical Solver

Due to the difficulty of solving high-order instances of bounded NP-complete combinatorial problems, many approximation and heuristic methods have been proposed in the literature. These methods, however, have unpredictable execution time. Therefore, for certain bounded NP-complete applications, where deadlines must be met, such methods are not a good choice and one may prefer to use an exhaustive search. For these cases, a new optical method which can provide a significantly better and guarantied solution-time is proposed in Refs. [2] and [3].

The proposed device is capable of solving bounded NP-complete problems such as the traveling salesman problem (TSP), the Hamiltonian path problem (HPP), etc. by checking all feasible possibilities orders of magnitude faster than a conventional computer. To do this, we use the VMM to perform a fast optical vector-by-matrix multiplication between a weight vector representing the problem weights and a binary matrix representing all feasible solutions. The multiplication product is a vector representing the final solutions of the problem. In the TSP for example, where the required solution is the shortest Hamiltonian tour connecting a certain set of given node coordinates, the multiplication is performed between a grayscale weight vector representing the weights between the TSP nodes and a binary matrix representing all feasible tours among the TSP nodes. The multiplication product is a length vector representing the TSP tour lengths by peaks of light with different intensities. Finding the shortest Hamiltonian tour can be performed by using an optical polynomial-time binary search which utilizes an optical threshold plate. On the other hand in the HPP, a decision whether there is a Hamiltonian path connecting two given nodes on the HPP graph is required. In the HPP, the binary matrix still represents all feasible paths (tours), but the weight vector is also binary. After performing the vector-by-matrix multiplication, any peak of light obtained in the output of the optical system means that a Hamiltonian path exists.

The advantage of the proposed method is that once the binary matrix is synthesized, all TSP and HPP instances of the same order (with the same number of nodes) can be solved optically by only changing the weight vector and performing the vector-by-matrix multiplication in an optical way. In addition, in Refs. [2] and [3] we have presented an efficient method to arrange the tours or paths in the binary matrix so that the binary matrix of N nodes contains the binary matrix of $N-1$ nodes. Therefore, once the binary matrix of N nodes is synthesized, all TSP and HPP instances containing N or fewer nodes can be solved by the VMM. In case the binary matrix contains more than 256×256 elements, the binary matrix is stored in the proposed device memory and then uploaded in a high rate to the SLM in parts in order to perform different portions of the vector-by-matrix multiplication each time. Note that in this case the SLM contains binary matrices (rather than 8 bit grayscale matrices). Hence, a dedicated special-purpose TSP/HPP VMM solver is expected to have performance that is 8 times faster than the performance of the general-purpose VMM presented in former sections.

5 Conclusions

New optical architecture for a super DSP co-processor that is based on of-the-shelf technology is presented. The architecture solves severe bottlenecks of the previous commercial electro-optical designs, gaining orders of magnitude better performance. The architecture supports principal- DSP primitives such as vector-by-matrix and matrix-by-matrix multiplications, convolution and correlation, discrete Fourier transform, L2 norm operations, addition, complex numbers and extended-precision arithmetic. These building blocks enable DSP-intensive applications, as well as an efficient bounded NP-complete problem solution. The device operational rate, in the order of at least 16-Tera integer operations per seconds, is a great opportunity for enhancing existing applications and introducing new computer applications, such as video compression and three-dimensional geometry engine. It is expected that developments in electro-optics technology will enable increasing the performance of this architecture, allowing more SED units and laser elements per device so that higher operational rate can be obtained.

Acknowledgments

This research was supported by Rita Altura Trust Chair in Computer Sciences.

References

1. Goren, A., Sariel, S., Levit, Y., Asaf, S., Liberman, B., Sender, T., Tzelnick, Y., Hefetz, H., Moses, E., Machal, V.: Vector-matrix multiplication. United States Patent No. 2004/0243657 A1 (2004)
2. Shaked, N.T., Messika, S., Dolev, S., Rosen, J.: Optical solution for bounded NP-complete problems. Applied Optics 46(5), 711–724 (2007)
3. Shaked, N.T., Tabib, T., Simon, G., Messika, S., Dolev, S., Rosen, J.: Optical binary-matrix synthesis for solving bounded NP-complete combinatorial problems. Optical Engineering 46(10), 108201, 1–11 (2007)
4. Dolev, S., Nir, Y.: Optical implementation of bounded non-deterministic Turing machines. US Patent No. 7,130,093 B2 (2006)
5. Dolev, S., Fitoussi, H.: The traveling beams optical solutions for bounded NP-complete problems. In: Crescenzi, P., Prencipe, G., Pucci, G. (eds.) FUN 2007. LNCS, vol. 4475, pp. 120–134. Springer, Heidelberg (2007)
6. Oppenheim, A.V., Schafer, R.W.: Discrete Time Signal Processing, 2nd edn. Prentice-Hall, Englewood Cliffs (1999)
7. Bier, J.: Byers Guide to DSP Processors, 2005 edn. Berkeley Design Technology Inc. (2006)
8. Anonymous, Texas Instruments C64 Performance Benchmarks, http://focus.ti.com/dsp/docs/dspplatformscontentaut tsp?sectionId=2& familyId-=477&tabId=496
9. Feitelson, D.G.: Optical Computing: A Survey for Computer Scientists. MIT Press, Cambridge (1988)
10. McAulay, D.: Optical Computer Architectures: The Application of Optical Concepts to Next Generation Computers. Wiley-Interscience, Chichester (1991)
11. Karim, M.A., Awwal, A.A.S.: Optical Computing: An Introduction. John Wiley & Sons, Chichester (1992)
12. Goodman, J.W.: Introduction to Fourier Optics. McGraw-Hill, New York (1996)
13. Tamir, D.E., Shaked, N.T., Wilson, P.J., Dolev, S.: Electro-optical DSP of Tera operations per second and beyond. Technical report #09-08, Department of Computer Science, Ben-Gurion University of the Negev, Beer Sheva, Israel (2008)
14. Gschwind, M.: Chip multiprocessing and the Cell broadband engine. IBM Research Report RC2931 (2006)
15. Anonymous, International Technology Roadmap for Semiconductors; Assembly and Packaging (ITRS 2005) (2005)
16. Wolf, J., Adams, J.: Packaging Roadmap Overview. The International Electronics Manufacturing Initiative (INEMI 2005) (2005)
17. Kim, J., Verbauwhede, I., Chang, M.F.: Design of an interconnect architecture and signaling technology for parallelism in communication. IEEE Transactions on Very Large Scale Integration (VLSI) Systems 15(8), 881–894 (2007)
18. Wentzlaff, D., Griffin, P., Hoffmann, H., Bao, L., Edwards, B., Ramey, C., Mattina, M., Miao, C., Brown, J.F., Agarwal, A.: On-chip interconnection architecture of the Tilera processor. IEEE Micro 27(5), 15–31 (2007)

19. Anonymous, Infiniband technology overview,
    http://www.infinibandta.org/about/
20. Hoskote, Y., Vangal, S., Singh, A., Borkar, N., Borkar, S.: A 5-GHz mesh interconnect for a Teraflops processor. IEEE Micro. 27(5), 51–61 (2007)
21. Wood, T.H., Carr, E.C., Burrus, C.A., Henry, J.E., Gossard, A.C., English, J.H.: High-speed 2X2 electrically driven spatial light modulator made with GaAs/AlGaAs multiple quantum wells (MQWs). Electronics Letters 23(17), 916–917 (1987)
22. Cormen, T.H., Leiserson, C.E., Rivest, R.L., Stein, R.C.: Introduction to Algorithms, 2nd edn. MIT Press, Cambridge (2001)
23. Anonymous, Basic local alignment search tool (BLAST),
    http://www.ncbi.nlm.nih.gov/blast/Blast.cgi

Parallel and Sequential Optical Computing[*]

.Damien Woods[1] and Thomas J. Naughton[2,3]

[1] Department of Computer Science and Artificial Intelligence,
University of Seville, Spain
d.woods@cs.ucc.ie
[2] Department of Computer Science
National University of Ireland, Maynooth, County Kildare, Ireland
[3] University of Oulu, RFMedia Laboratory, Oulu Southern Institute, Vierimaantie 5,
84100 Ylivieska, Finland
tomn@cs.nuim.ie

Abstract. We present a number of computational complexity results for
an optical model of computation called the continuous space machine.
We also describe an implementation for an optical computing algorithm
that can be easily defined within the model. Our optical model is designed to model a wide class of optical computers, such as matrix vector
multipliers and pattern recognition architectures. It is known that the
model solves intractable PSPACE problems in polynomial time, and NC
problems in polylogarithmic time. Both of these results use large spatial
resolution (number of pixels). Here we look at what happens when we
have constant spatial resolution. It turns out that we obtain similar results by exploiting other resources, such as dynamic range and amplitude
resolution. However, with certain other restrictions we essentially have
a sequential device. Thus we are exploring the border between parallel
and sequential computation in optical computing. We describe an optical
architecture for the unordered search problem of finding a one in a list of
zeros. We argue that our algorithm scales well, and is relatively straightforward to implement. This problem is easily parallelisable and is from
the class NC. We go on to argue that the optical computing community
should focus their attention on problems within P (and especially NC),
rather than developing systems for tackling intractable problems.

1 Introduction

Over the years, optical computers were designed and built to emulate conventional microprocessors (digital optical computing), and for image processing over
continuous wavefronts (analog optical computing). Here we are interested in the
latter class: optical computers that store data as images. Numerous physical
implementations exist and example applications include fast pattern recognition
and matrix-vector algebra. There have been much resources devoted to designs,

---

[*] DW acknowledges support from Junta de Andalucía grant TIC-581. TN acknowledges support from a Marie Curie Fellowship through the European Commission
Framework Programme 6.

S. Dolev, T. Haist, and M. Oltean (Eds.): OSC 2008, LNCS 5172, pp. 70–86, 2008.

implementations and algorithms for such optical information processing architectures (for example see [1,8,11,?,29] and their references).

We investigate the computational complexity of a model of computation that is inspired by such optical computers. The model is relatively new and is called the continuous space machine (CSM). The model was originally proposed by Naughton [17,18]. The CSM computes in discrete timesteps over a number of two-dimensional images of fixed size and arbitrary spatial resolution. The data and program are stored as images. The (constant time) operations on images include Fourier transformation, multiplication, addition, thresholding, copying and scaling. We analyse the model in terms of seven complexity measures inspired by real-world resources.

For the original [18] CSM definition, it was shown [17] that the CSM can simulate Turing machines (this was a sequential simulation). A less restricted CSM definition [20,35] was shown to be too general for proving reasonable upper bounds on its computational power [33], so in this paper we mostly focus on computational complexity results for a restricted CSM called the C_2-CSM.

In Section 2 we recall the definition of the model, including a number of optically-inspired complexity measures [35]. In Section 2.5 we describe a number of known computational complexity results for the model, including characterisations of PSPACE and NC. These results were shown a few years ago [30,34] (later improved [32]), and were the first to prove that optical computers were capable of solving NP-complete (and other intractable) problems in polynomial time. Of course, these results make use of exponential space-like resources. In particular, these algorithms used exponential *spatial resolution* (number of pixels). Since we have a clear model definition, including definitions of relevant optical resources, it is relatively easy to analyse CSM algorithms to determine their resource usage. Recently, Shaked et al. [25,26,27] have designed an optical system for solving the NP-hard travelling salesman problem in polynomial time. Their algorithm can be seen as a special case of our results. Interestingly, they give both implementations and simulations. As we argue below, we believe that tackling intractable problems is probably not going to really highlight any advantages of optics over digital electronic systems. As a step in another direction, we have shown that if we restrict ourselves to using polylogarithmic time, and polynomial space-like resources, then parallel optical systems can solve exactly those problems that lie in the (parallel) class NC.

In Section 3 we present a number of new results for our model. In particular we look at what happens when spatial resolution is constant. Parallel optical algorithms and experimental setups usually exploit the fact that we can operate over many pixels in constant time. However, we show that even with a constant number of pixels we can solve problems in (and characterise) presumed intractable classes such as PSPACE, in polynomial time. In this case we make exponential usage of other resources, namely amplitude resolution and dynamic range. We argue that this is an even more unrealistic method of optical computing than using exponential numbers of pixels. We go on to show that if we disallow image multiplication, restrict to polynomial numbers of pixels and/or images, but put no restrictions on the other resources, then in polynomial time the model characterises P.

This results lead us to suggest of a new direction for optical algorithm designers. Rather than trying to solve intractable problems, perhaps the community should focus its attention on problems that are known to be easily parallelisable, for example those in NC. Of course, these problems are polynomial time solvable on sequential machines. However, using our NC characterisations one can see that optics has the potential to solve such problems exponentially faster than sequential computers. Also, due to relatively low communication costs and high fan-in, optics has the potential to out-perform parallel digital electronic architectures. Perhaps such benefits of optics will be seen where very large datasets and input instances are concerned. We give evidence for this opinion by citing existing optical algorithms, as well as the following result in this paper.

We design an optoelectronic implementation for the unordered search problem of finding a single one in a list of zeros. Of course, this problem can be sequentially solved in $n-1$ steps. Our algorithm works in $O(\log n)$ time but, most importantly, we get this low time overhead on an optical set-up that scales well (uses at most n pixels), and is relatively straightforward to build. As we discuss in Section 4.1, this problem is contained in some of the lowest classes within NC.

2   CSM and $\mathcal{C}_2$-CSM

We begin by describing the model in its most general sense, this brief overview is not intended to be complete and more details are to be found in [30].

2.1   CSM

A complex-valued image (or simply, image) is a function $f : [0, 1) \times [0, 1) \to \mathbb{C}$, where $[0, 1)$ is the half-open real unit interval. We let $\mathcal{I}$ denote the set of complex-valued images. Let $\mathbb{N}^+ = \{1, 2, 3, \ldots\}$, $\mathbb{N} = \mathbb{N}^+ \cup \{0\}$, and for a given CSM M let $\mathcal{N}$ be a countable set of images that encode M's addresses. An address is an element of $\mathbb{N} \times \mathbb{N}$.

Definition 1 (CSM). *A CSM is a quintuple $M = (\mathfrak{E}, L, I, P, O)$, where*

$\mathfrak{E} : \mathbb{N} \to \mathcal{N}$ *is the address encoding function,*
$L = ((s_\xi, s_\eta), (a_\xi, a_\eta), (b_\xi, b_\eta))$ *are the addresses: sta, a and b, where $a \neq b$,*
I *and O are finite sets of input and output addresses, respectively,*
$P = \{(\zeta_1, p_{1_\xi}, p_{1_\eta}), \ldots, (\zeta_r, p_{r_\xi}, p_{r_\eta})\}$ *are the r programming symbols ζ_j and*
    their addresses where $\zeta_j \in (\{h, v, *, \cdot, +, \rho, st, ld, br, hlt\} \cup \mathcal{N}) \subset \mathcal{I}$.
Each address is an element from $\{0, \ldots, \Xi-1\} \times \{0, \ldots, \mathcal{Y}-1\}$ where $\Xi, \mathcal{Y} \in \mathbb{N}^+$.

Addresses whose contents are not specified by P in a CSM definition are assumed to contain the constant image $f(x, y) = 0$. We interpret this definition to mean that M is (initially) defined on a grid of images bounded by the constants Ξ and $\mathcal{Y}$, in the horizontal and vertical directions respectively. The grid of images

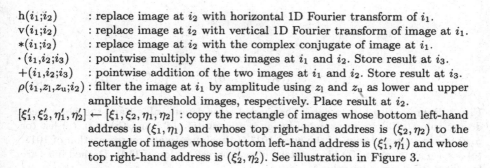

$\mathrm{h}(i_1;i_2)$: replace image at i_2 with horizontal 1D Fourier transform of i_1.

$\mathrm{v}(i_1;i_2)$: replace image at i_2 with vertical 1D Fourier transform of image at i_1.

$*(i_1;i_2)$: replace image at i_2 with the complex conjugate of image at i_1.

$\cdot(i_1,i_2;i_3)$: pointwise multiply the two images at i_1 and i_2. Store result at i_3.

$+(i_1,i_2;i_3)$: pointwise addition of the two images at i_1 and i_2. Store result at i_3.

$\rho(i_1,z_\mathrm{l},z_\mathrm{u};i_2)$: filter the image at i_1 by amplitude using z_l and z_u as lower and upper amplitude threshold images, respectively. Place result at i_2.

$[\xi_1',\xi_2',\eta_1',\eta_2'] \leftarrow [\xi_1,\xi_2,\eta_1,\eta_2]$: copy the rectangle of images whose bottom left-hand address is (ξ_1,η_1) and whose top right-hand address is (ξ_2,η_2) to the rectangle of images whose bottom left-hand address is (ξ_1',η_1') and whose top right-hand address is (ξ_2',η_2'). See illustration in Figure 3.

Fig. 1. CSM high-level programming language instructions. In these instructions $i, z_\mathrm{l}, z_\mathrm{u} \in \mathbb{N} \times \mathbb{N}$ are image addresses and $\xi, \eta \in \mathbb{N}$. The control flow instructions are described in the main text.

may grow in size as the computation progresses. Address sta is the start location for the program so the programmer should write the first program instruction (beginning) at sta. Addresses a and b define special images that are frequently used by some program instructions.

In our grid notation the first and second elements of an address tuple refer to the horizontal and vertical axes of the grid respectively, and image $(0,0)$ is located at the lower left-hand corner of the grid. The images have the same orientation as the grid. For example the value $f(0,0)$ is located at the lower left-hand corner of the image f.

In Definition 1 the tuple P specifies the CSM program using programming symbol images ζ_j that are from the (low-level) CSM programing language [30,35]. We refrain from giving a description of this programming language and instead describe a less cumbersome high-level language [30]. Figure 1 gives the basic instructions of this high-level language. The copy instruction is illustrated in Figure 3. There are also **if/else** and **while** control flow instructions with conditions of the form $(f_\psi == f_\phi)$ where f_ψ and f_ϕ are *binary symbol images* (see Figures 2(a) and 2(b)).

The function $\mathfrak{E}$ is specified by the programmer and is used to map addresses to image pairs. This enables the programmer to choose her own address encoding scheme. We typically don't want $\mathfrak{E}$ to hide complicated behaviour thus the computational power of this function should be somewhat restricted. Thus we insist that for a given M there is an *address encoding function* $\mathfrak{E} : \mathbb{N} \to \mathcal{N}$ such that $\mathfrak{E}$ is Turing machine decidable, under some *reasonable* representation of images as words. For example, we put a restriction of logspace computability on $\mathfrak{E}$ in Definition 7 below. Configurations are defined in a straightforward way as a tuple $\langle c, e \rangle$ where c is an address called the control and e represents the grid contents.

2.2   Complexity Measures

Next we define some CSM complexity measures. All resource bounding functions map from $\mathbb{N}$ into $\mathbb{N}$ and are assumed to have the usual properties [2].

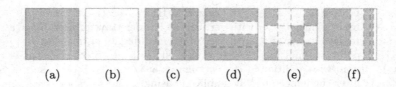

(a)   (b)   (c)   (d)   (e)   (f)

Fig. 2. Representing binary data. The shaded areas denote value 1 and the white areas denote value 0. (a) Binary symbol image representation of 1 and (b) of 0, (c) list (or row) image representation of the word 1011, (d) column image representation of 1011, (e) 3×4 matrix image, (f) binary stack image representation of 1101. Dashed lines are for illustration purposes only.

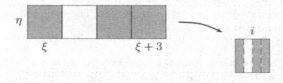

Fig. 3. Illustration of the instruction $i \leftarrow [\xi, \xi + 3, \eta, \eta]$ that copies four images to a single image that is denoted i.

Definition 2. *The* TIME *complexity of a CSM M is the number of configurations in the computation sequence of M, beginning with the initial configuration and ending with the first final configuration.*

Definition 3. *The* GRID *complexity of a CSM M is the minimum number of images, arranged in a rectangular grid, for M to compute correctly on all inputs.*

Let $S : \mathcal{I} \times (\mathbb{N} \times \mathbb{N}) \rightarrow \mathcal{I}$, where $S(f(x,y), (\Phi, \Psi))$ is a raster image, with $\Phi\Psi$ constant-valued pixels arranged in Φ columns and Ψ rows, that approximates $f(x,y)$. If we choose a reasonable and realistic S then the details of S are not important.

Definition 4. *The* SPATIALRES *complexity of a CSM M is the minimum $\Phi\Psi$ such that if each image $f(x,y)$ in the computation of M is replaced with $S(f(x,y), (\Phi, \Psi))$ then M computes correctly on all inputs.*

One can think of SPATIALRES as a measure of the number of pixels needed during a computation. In optical image processing terms, and given the fixed size of our images, SPATIALRES corresponds to the space-bandwidth product of a detector or spatial light modulator.

Definition 5. *The* DYRANGE *complexity of a CSM M is the ceiling of the maximum of all the amplitude values stored in all of M's images during M's computation.*

In optical processing terms DYRANGE corresponds to the dynamic range of a signal.

We also use complexity measures called AMPLRES, PHASERES and FREQ [30,35]. Roughly speaking, the AMPLRES of a CSM M is the number of discrete, evenly spaced, amplitude values per unit amplitude of the complex numbers in the range of M's images. The PHASERES of M is the total number (per 2π) of discrete evenly spaced phase values in the range of M's images. FREQ is a measure of the optical frequency of M's images [35].

Often we wish to make analogies between space on some well-known model and CSM 'space-like' resources. Thus we define the following convenient term.

Definition 6. *The* SPACE *complexity of a CSM M is the product of all of M's complexity measures except* TIME.

2.3 Representing Data as Images

There are many ways to represent data as images. Here we mention some data representations that we have used in previous results. Figures 2(a) and 2(b) are the binary symbol image representations of 1 and 0 respectively. These images have an everywhere constant value of 1 and 0 respectively, and both have SPATIALRES of 1. The row and column image representations of the word 1011 are respectively given in Figures 2(c) and 2(d). These row and column images both have SPATIALRES of 4. In the matrix image representation in Figure 2(e), the first matrix element is represented at the top left corner and elements are ordered in the usual matrix way. This 3×4 matrix image has SPATIALRES of 12. Finally, the binary stack image representation, which has exponential SPATIALRES of 16, is given in Figure 2(f).

Figure 3 shows how we might form a list image by copying four images to one in a single timestep. All of the above mentioned images have DYRANGE, AMPLRES, and PHASERES of 1.

2.4 C_2-CSM

Motivated by a desire to apply standard complexity theory tools to the model, we defined [30,33] the C_2-CSM, a restricted class of CSM.

Definition 7 (C_2-CSM). *A C_2-CSM is a CSM whose computation* TIME *is defined for $t \in \{1, 2, \ldots, T(n)\}$ and has the following restrictions:*

- *For all* TIME *t both* AMPLRES *and* PHASERES *have constant value of 2.*
- *For all* TIME *t each of* GRID, SPATIALRES *and* DYRANGE *is $O(2^t)$ and* SPACE *is redefined to be the product of all complexity measures except* TIME *and* FREQ.
- *Operations h and v compute the discrete Fourier transform (DFT) in the horizontal and vertical directions respectively.*
- *Given some* reasonable *binary word representation of the set of addresses $\mathcal{N}$, the address encoding function $\mathfrak{E} : \mathbb{N} \to \mathcal{N}$ is decidable by a logspace Turing machine.*

Let us discuss these restrictions. The restrictions on AMPLRES and PHASERES imply that C_2-CSM images are of the form $f : [0, 1) \times [0, 1) \rightarrow \{0, \pm\frac{1}{2}, \pm1, \pm\frac{3}{2}, \ldots\}$. We have replaced the Fourier transform with the DFT [5], this essentially means that FREQ is now solely dependent on SPATIALRES; hence FREQ is not an interesting complexity measure for C_2-CSMs and we do not analyse C_2-CSMs in terms of FREQ complexity [30,33]. Restricting the growth of SPACE is not unique to our model, such restrictions are to be found elsewhere [10,21,22]. The condition on the address encoding function $\mathfrak{E}$ amounts to enforcing uniformity (we do not wish to use $\mathfrak{E}$ as a powerful oracle).

In this paper we prove results for variants (generalisations and restrictions) on the C_2-CSM model. If we are not stating results for the C_2-CSM itself, then we always specify the exact model that we are using.

2.5   Some Existing C_2-CSM Complexity Results

We have given lower bounds on the computational power of the C_2-CSM by showing that it is at least as powerful as models that verify the parallel computation thesis [30,32,34]. This thesis [7,9] states that parallel time corresponds, within a polynomial, to sequential space for reasonable parallel models. See, for example, [12,14,21,28] for details. Let $S(n)$ be a space bound that is $\Omega(\log n)$. The languages accepted by nondeterministic Turing machines in $S(n)$ space are accepted by C_2-CSMs computing in polynomial TIME $O(S^2(n))$ (see [32] for this result, which improves on the version in [30,34]):

Theorem 1. NSPACE$(S(n)) \subseteq C_2$-CSM–TIME$(O(S^2(n)))$

For example, polynomial TIME C_2-CSMs accept the PSPACE languages[1]. Of course any polynomial TIME C_2-CSM algorithm that we could presently write to solve PSPACE-complete, or NP-complete, problems would require exponential SPACE. Theorem 1 is established using an implementation of a well-known transitive closure algorithm on the C_2-CSM. Using this result, we also find that C_2-CSMs that simultaneously use polynomial SPACE and polylogarithmic TIME accept the class NC [30,34].

Corollary 1. NC $\subseteq C_2$-CSM–SPACE, TIME$(n^{O(1)}, \log^{O(1)} n)$

We have also given the other of the two inclusions that are necessary in order to verify the parallel computation thesis: C_2-CSMs computing in TIME $T(n)$ are no more powerful than $O(T^2(n))$ space bounded deterministic Turing machines [30,31].

Theorem 2. C_2-CSM-TIME$(T(n)) \subseteq$ DSPACE$(O(T^2(n)))$

Via the proof of Theorem 2, we get another result. C_2-CSMs that simultaneously use polynomial SPACE and polylogarithmic TIME accept at most NC [30,31].

---

[1] PSPACE is a well-known class of problems that are solvable by Turing machines that use space polynomial in input length n. This class contains NP, since a polynomial space bounded Turing machine can simulate, in turn, each of the exponentially many possible computation paths of a nondeterministic polynomial time Turing machine.

Corollary 2. C_2-CSM-SPACE, TIME($n^{O(1)}, \log^{O(1)} n$) $\subseteq$ NC

The latter two inclusions are established via C_2-CSM simulation by logspace uniform circuits of size and depth polynomial in SPACE and TIME respectively. Thus C_2-CSMs that simultaneously use both polynomial SPACE and polylogarithmic TIME characterise NC.

3   Parallel and Sequential C_2-CSM Computation

As we have seen in the previous section, a number of computational complexity results for the C_2-CSM have shown that the model is capable of parallel processing in much the same way as models that verify the parallel computation thesis, and models that are known to characterise the parallel class NC. To date, these results strongly depended on their use of non-constant SPATIALRES. The algorithms exploit the ability of optical computers, and the CSM in particular, to operate on large numbers of pixels in parallel. But what happens when we do not have arbitrary numbers of pixels? If allow images to have only a constant number of pixels then we need to find new CSM algorithms. It turns out that that such machines characterise PSPACE.

Theorem 3. PSPACE *is characterised by* C_2-CSMs *that are restricted to use polynomial* TIME $T = O(n^k)$, SPATIALRES $O(1)$, GRID $O(1)$, *and generalised to use* AMPLRES $O(2^{2^T})$, DYRANGE $O(2^{2^T})$.

Proof. The PSPACE upper bound comes directly from a minor extension to the proof of Theorem 2, sketched as follows. The proof of Theorem 2 showed that C_2-CSMs that run in polynomial TIME $T = O(n^k)$, are simulated by circuits of polynomial depth $O(T^2)$ and size exponential in T, and it remains to be shown that our AMPLRES and DYRANGE generalisations do not affect these circuit bounds by more than a polynomial factor. In the previous proof [30,31] DYRANGE was $O(2^T)$, in accordance with the usual C_2-CSM definition. Thus, in the circuit simulation, images values $x \in \{1, \ldots, O(2^T)\} \subseteq \mathbb{N}$ were represented by binary words $\hat{x}$ of length $|\hat{x}| = O(T)$. We directly apply the previous construction to represent values $x \in \{1, \ldots, O(2^{2^T})\}$ as words of length $|\hat{x}| = O(2^T)$. Since the circuits are already of size exponential in T, this modification only increases circuit size by a polynomial factor in the existing size. Also, the circuit simulation algorithms experience at most a polynomial factor increase in their depth. A similar argument works for AMPLRES (even though in the previous proof AMPLRES was $O(1)$). Here we are using constant SPATIALRES and GRID (as opposed to $O(2^T)$ for the previous proof), so circuit size and depth are each decreased by a polynomial factor in their respective previous values. We omit the details.

For the lower bound we use the results of Schönhage [24] and Bertoni et al. [4] which show that PSPACE is characterised by RAMs augmented with integer addition, multiplication, and left shift instructions, that run in time that is polynomial in input length n. We show how to simulate such an augmented RAM with a C_2-CSM that has time overhead that is polynomial in RAM time.

The numerical value $x \in \mathbb{N}$ of the binary word in a RAM register is stored as an image, with a single pixel, of value x. The RAM uses a constant (independent of input length n) number of registers, and therefore the C_2-CSM uses a constant number of images. The addition and multiplication RAM operations are trivially simulated in constant time by C_2-CSM addition and multiplication instructions.

The RAM shift instruction $x \leftarrow y$ takes a register x containing a binary value and shifts it to the right by an amount stored in another binary register y (Schönhage defines the shift instruction as $\lfloor x/2^y \rfloor$). In the C_2-CSM this can be simulated (using multiplication and addition) by $(x \cdot 1/2^y) + (-1 \cdot x')$ where x' is the result of (thresholding and multiplication) $\rho(x \cdot 1/2^y, 1/2^y, 1; x')$, and the value $1/2^y$ is computed by repeated multiplication in $O(\log y)$ steps.

The C_2-CSM algorithm uses AMPLRES and DYRANGE that are exponential in the space used by the RAM and TIME polynomial in the time of the RAM. All other resources are constant.   □

So by treating images as registers and generating exponentially large, and exponentially small, values we can solve seemingly intractable problems. Of course this kind of CSM is quite unrealistic from the point of view of optical implementations. In particular, accurate multiplication of such values in optics is difficult to implement. Some systems have up to a few hundred distinct amplitude levels [8] (for example 8 bits when we have 256×256 pixels [15], although higher accuracy is possible when we have a single pixel[2]). Therefore, one could argue that this kind of multiplication is quite unrealistic. To restrict the model we could replace arbitrary multiplication, by multiplication by constants, which can be easily simulated by a constant number of additions. If we disallow multiplication in this way, we characterise P.

Theorem 4. C_2-CSM*s without multiplication, that compute in polynomial* TIME, *polynomial* GRID $O(n^k)$, *and* SPATIALRES $O(1)$, *characterise* P.

Proof (Sketch). For the P lower bound assume that we wish to simulate a deterministic Turing machine with one-way infinite tapes. Each tape is represented as a row of images, one for each tape cell. We store a pointer to the current tape head position as an image address. Then it is a straightforward matter to convert the Turing machine program to CSM program, where a left (right) move corresponds to decrementing (incrementing) the address pointer. Reading and writing to the tape is simulated by copying images. CSM branching instructions simulate branching in the Turing machine program. The CSM runs in TIME that is linear in Turing machine time.

For the P upper bound we assume some representation of images as binary words (such as the representation given above, or in [30,31]), and apply a simple inductive argument. The initial configuration of our restricted C_2-CSM is encoded by a binary word of length polynomial in the input length n. Assume at

---

[2] Inexpensive off-the-shelf single point intensity detectors have intensity resolutions of at least 24 bits (see, for example, the specifications for silicon-based point detectors and optical power meters from popular manufacturers such as www.mellesgriot.com, www.newport.com, and www.thorlabs.com).

$\mathcal{C}_2$-CSM TIME t that the binary word encoding the configuration is of polynomial length. For each pixel, addition of two images leads to an increase of at most one bit per pixel under our representation, and can be simulated in polynomial time on a Turing machine. The DFT over a finite field is computable in polynomial time and is defined in such a way that it does not increase the number of pixels in an image. Also, its input and output values are from the same set, therefore the upper bounds on the other space-like resources are unaffected by the DFT. Copying (up to) a polynomial number of encoded images can be computed in polynomial time. It is straightforward to simulate complex conjugation and thresholding in linear time. □

The first proof of universality for the CSM was a simulation of Turing machines that used SPACE that is exponential in Turing machine space [17]. Specifically, it used constant GRID and exponential SPATIALRES. The previous theorem improves the SPACE bound to linear, by using linear GRID and only constant SPATIALRES.

If we take the previous restricted $\mathcal{C}_2$-CSM, and restrict further to allow only constant GRID, but allow ourselves polynomial SPATIALRES, then we also characterise P.

Theorem 5. *CSMs without multiplication, that compute in polynomial* TIME, *polynomial* SPATIALRES $O(n^k)$, *and* GRID $O(1)$, *characterise P.*

Proof (Sketch). Here we are considering a $\mathcal{C}_2$-CSM model that is similar to the model in Theorem 4; we are swapping GRID for SPATIALRES. Hence a very similar technique can be used to show the P upper bound, so we omit the details.

For the lower bound, we store each one-way, polynomial $p(n)$ length, binary, Turing machine tape as a binary list image. We store the current tape head position $i \in \{1, \ldots, p(n)\}$ as a binary list image that represents i in binary. Then, to extract the bit stored at position i, we can use a $O(\log p(n))$ TIME binary search algorithm (split tape image in two, if $i \leqslant p(n)/2$ then take the left image, otherwise take the right, let $p(n) := p(n)/2$ and repeat). This technique, along with suitable masks, can also be applied to write to the tape. The Turing machine program is simulated using $\mathcal{C}_2$-CSM branching instructions. □

Theorems 4 and 5 give conditions under which our optical model essentially looses its parallel abilities and acts like a standard sequential Turing machine.

4   Implementation of an Unordered Search Algorithm

We provide a design for an optoelectronic implementation of a binary search algorithm that can be applied to unordered lists. Consider an unordered list of n elements. For a given property P, the list could be represented by an n-tuple of bits, where the bit key for each element denotes whether or not that element satisfies P. If, for a particular P, only one element in the list satisfies P, the problem of finding its index becomes one of searching an unordered binary list for a single 1. The problem is defined formally as follows.

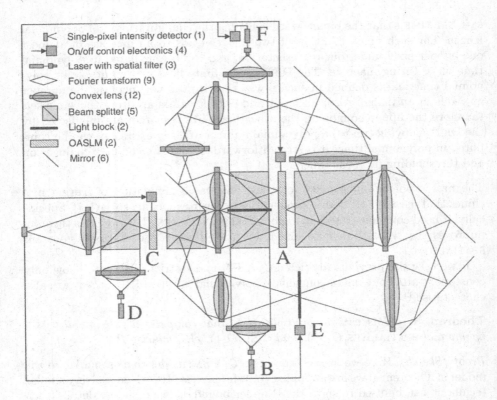

Fig. 4. Optical apparatus to perform a binary search for a single 1 in a bitstream of 0s. The legend explains the optical components and the number of them required. The labels A-F are explained in the text. OASLM: optically-addressed spatial light modulator.

Definition 8 (Needle in haystack problem). *Let* $L = \{w : w \in 0^*10^*\}$. *Let* $w \in L$ *be written as* $w = w_0w_1\ldots w_{n-1}$ *where* $w_i \in \{0,1\}$. *Given such a* w, *the needle in haystack problem asks what is the index of the symbol 1 in* w. *The solution to the needle in haystack problem for a given* w *is the index* i, *expressed in binary, where* $w_i = 1$.

This problem was posed by Grover in [13]. His quantum computer algorithm requires $O(\sqrt{n})$ comparison operations on average. Bennett *et al.* [3] have shown the work of Grover is optimal up to a multiplicative constant, and that in fact any quantum mechanical system will require $\Omega(\sqrt{n})$ comparisons. It is not difficult to see that algorithms for sequential models of computation require $\Theta(n)$ comparisons in the worst case to solve the problem. We present an algorithm that requires $O(\log_2 n)$ comparisons, with a model of computation that has promising future implementation prospects.

Our search algorithm is quite simple. A single bright point is somewhere in an otherwise dark image. If we block one half of the image we can tell in a single step if the other half contains the bright point or not. This forms the basis of a binary search algorithm to determine the precise location of the bright point.

Theorem 6 ([35]). *There exists a CSM that solves the needle in haystack problem in $\Theta(\log_2 n)$ comparisons for a list of length n, where $n = 2^k, k \in \mathbb{N}, k \geq 1$.*

The CSM instance that performs this computation is given elsewhere [35], as are details of a counter to determine when the computation has finished and details of how the next most significant bit of the address is built up in an image at each step. We explain the main loop of the CSM algorithm. The binary input list w can be represented as a binary list image as illustrated in Fig. 2(c), or more simply by using a single point of light instead of a vertical stripe to denote value 1. Therefore, w would be represented by a small bright spot (a high amplitude peak) in an otherwise black image, where the position of the peak denotes the location of the 1 in w. During the first iteration of the loop, w is divided equally into two images (a left-hand image and a right-hand image). The nonzero peak will be either in the left-hand image or the right-hand image. In order to determine which image contains the peak in a constant number of steps, the left-hand image is Fourier transformed, squared, and Fourier transformed again. This effectively moves the nonzero peak (wherever it was originally) to the centre of the image, where it can be easily compared to a template (using a single conditional branching instruction in the CSM). If the left-hand image contains a nonzero amplitude at its centre, then the left-hand image contained the peak. In this case, the right-hand image is discarded, and the most significant bit of the address of 1 in w is 0. Otherwise, the right-hand image contained the peak, the left-hand image is discarded, and the most significant bit of the address is 1. For the next iteration, the remainder of the list is divided equally into two images and the computation proceeds according to this binary search procedure.

A schematic for an optoelectronic implementation is shown in Fig. 4. At the start of the computation, optically addressed spatial light modulator (OASLM) A is initialized to display the input list. One half of the input is read out of the OASLM using illumination B and a beam splitter in standard configuration, and is transformed by a Fourier lens so that its Fourier transform falls on OASLM C. The act of detection within C squares the image, and it is read out and Fourier transformed again using illumination D. A single point detector is centred on the optical axis – exactly where light would fall if there was a bright spot anywhere in the left half of the list on OASLM A. If light is detected, light block E is opened to allow the left half of the list to be copied to the full extent of A, otherwise illumination F is switched on to allow the right half of the list to be copied to the full extent of A. The act of detection within A will itself square the image but this is no concern because it is intended that the list would have constant phase everywhere. In fact, the response of A could be configured to nonlinearly transform the intensities in the image, to suppress any background noise and enhancing the bright spot, thus avoiding the propagation of errors which is the overriding problem with numerical calculations implemented in analog optics [16]. The left half of the remainder of the list is now ready to be Fourier transformed itself onto C, for the next step in the computation. For simplicity, at each step we let the next bit of the address to be recorded at the detector electronics.

To more rigorously verify the operation of the apparatus in Fig. 4 from an optical engineering standpoint, the following steps in the process are stated explicitly.

Step 1: Ensure F is off, E closed, B on, D on, C cleared, input displayed on A.
Step 2: (Detector will sense the presence or absence of light.) Put A into write mode. If light is sensed, record address bit of 0 and open E for an instant, otherwise record an address bit of 1 and switch F on for an instant. Take A out of write mode. Clear C.
Step 3: Go to step 2.

All that is required to augment the design is a counter to determine when to halt (which can also be performed electronically) and a method of initialising OASLM A with the input (which can be performed by replacing the upper-right mirror with a beam splitter). This is by no means a definitive implementation of the operation, but conceptually it is very simple, and as a design it is straightforward to implement. The most difficult implementation issues concern ensuring that light close to the centre of A is appropriately partitioned by the pair of side-by-side Fourier lenses, and ensuring that the feedback paths (the two paths from A to itself) are not unduly affected by lens abberations. Ultimately, the spatial resolution of the input images (and so the size of the list inputs) is limited by the finite aperture size of the lenses. Practically, it would be desirable to configure both B and D to be pulsed in the same way as F, although this adds to the control burden. It would be possible to replace the two 4-f feedback paths with 2-f feedback paths (thereby removing two lenses from each path) if one took note that the list would be reversed at each alternate step. Further, each pair of Fourier lenses in the upper feedback arm could be replaced by a single lens in imaging configuration if one ignores the phase errors – Fourier transforming lenses are used exclusively in this design to ease detailed verification of the apparatus by the reader. Imaging lenses would also allow reduction in size of the largest of the mirrors and beamsplitter in the design. Furthermore, passive beamsplitters and planar mirrors are specified here to maintain the quality of the optical wavefronts at reasonable notional financial cost; instead employing active beam splitter technology and curved mirrors would reduce the number of components further while admitting their own disadvantages. Finally, cylindrical lenses could be used rather than spherical lenses because Fourier transformation in one dimension only is required.

4.1   Complexity of the Unordered Search Problem

It is possible to give an AC^0 circuit family to solve the Needle in haystack problem. In fact, is is possible to give constant time CSM, or C_2-CSM, algorithms to solve the problem. However, although fast, we felt than any such CSM algorithm that we could think of was more difficult to implement optically than the above algorithm. For example, one can consider a CSM algorithm that encodes the values $1, \ldots n$ at addresses $a_1, \ldots, a_n$, respectively. Next we assume an ordering on the n possible inputs that corresponds to the ordering in the addressing scheme. Using such an input, the machine would simply branch to the address at the input's value, and output the image at the ith address. The algorithm runs in constant TIME, linear GRID, and all other resources are constant. Although simple to describe, the use of addressing would complicate the algorithm's implementation.

5   A New Direction for Optical Algorithm Designers?

Nature-inspired systems that apparently solve NP-hard problems in polynomial time, while using an exponential amount of some other resource(s), have been around for many years. So the existence of massively parallel optical systems for NP-hard problems should not really suprise the reader.

One could argue that it is interesting to know the computational abilities, limitations, and resource trade-offs of such optical architectures, as well as to find particular (tractable or intractable) problems which are particularly suited to optical algorithms. However, "algorithms" that use exponential space-like resources (such as number of pixels, number of images, number of amplitude levels, etc.) are not realistic to implement for large input instances. What happens to highly parallel optical architectures if add the requirement that the amount of space-like resources are bounded in some reasonable way? We could, for example, stipulate that the optical machine use no more than a polynomially bounded amount of space-like resources. If the machine runs in polynomial time, then it is not difficult to see that it characterises P (by characterise we mean that the model solves exactly those problems in P), for a wide range of reasonable parallel and sequential optical models (see Section 3). Many argue that the reason for using parallel architectures is to speed-up computations. Asking for an exponential speed-up motivates the complexity class NC. The class NC can be thought of as those problems in P that can be solved exponentially faster on parallel computers than on sequential computers. Thus, NC is contained in P and it is an major open question whether this containment is strict: it is widely conjectured that this is indeed the case [12].

How does this relate to optics? As discussed in Section 2.5, a wide range of optical computers that run for at most polylogarithmic time, and use at most polynomial space-like resources, solve exactly NC [30,31,32,34]. In effect this means that we have an algorithmic method (in other words, a compiler) to convert existing NC algorithms into optical algorithms that use similar amounts of resources.

From the practical point of view, perhaps we can use these kinds of results to find problems within NC, where optical architectures can be shown to excel. Obvious examples for which this is already known are matrix-vector multiplication (which lies in NC^2), or Boolean matrix multiplication (which is in NC^1).[3] Another example is the NC^1 unordered search problem given in Section 4. Another closely related idea is to exploit the potential unbounded fan-in of optics to compute problems in the AC, and TC, (parallel) circuit classes. These are defined similarly to NC circuits except we allow unbounded fan-in gates, and threshold gates, respectively. The results of Reif and Tyagi [23], and Caulfield's observation on the benefits of unbounded fan-in [6], can be interpreted as exploiting this important and efficient aspect of optics.

There is scope for further work here, on the continuous space machine in particular, in order to find exact characterisations, or as close as possible for NC^k for given k. Or even better, to find exact characterisations of the AC^k or TC^k classes of problems.

References

1. Arsenault, H.H., Sheng, Y.: An Introduction to Optics in Computers. Tutorial Texts in Optical Engineering, vol. TT8. SPIE Press, Bellingham (1992)
2. Balcázar, J.L., Díaz, J., Gabarró, J.: Structural complexity II. EATCS Monographs on Theoretical Computer Science, vol. 22. Springer, Berlin (1988)
3. Bennett, C.H., Bernstein, E., Brassard, G., Vazirani, U.: Strengths and weaknesses of quantum computing. SIAM Journal on Computing 26(5), 1510–1523 (1997)
4. Bertoni, A., Mauri, G., Sabadini, N.: A characterization of the class of functions computable in polynomial time on random access machines. In: STOC, Milwaukee, Wisconsin, May 1981, pp. 168–176. ACM Press, New York (1981)
5. Bracewell, R.N.: The Fourier transform and its applications, 2nd edn. Electrical and electronic engineering series. McGraw-Hill, New York (1978)
6. Caulfield, H.J.: Space-time complexity in optical computing. Multidimensional Systems and Signal Processing 2(4), 373–378 (1991); Special issue on optical signal processing
7. Chandra, A.K., Stockmeyer, L.J.: Alternation. In: 17th annual symposium on Foundations of Computer Science, Houston, Texas, October 1976, pp. 98–108. IEEE, Los Alamitos (1976)
8. Feitelson, D.G.: Optical Computing: A survey for computer scientists. MIT Press, Cambridge (1988)
9. Goldschlager, L.M.: Synchronous parallel computation. PhD thesis, University of Toronto, Computer Science Department (December 1977)
10. Goldschlager, L.M.: A universal interconnection pattern for parallel computers. Journal of the ACM 29(4), 1073–1086 (1982)
11. Goodman, J.W.: Introduction to Fourier optics, 2nd edn. McGraw-Hill, New York (1996)
12. Greenlaw, R., Hoover, H.J., Ruzzo, W.L.: Limits to parallel computation: P-completeness theory. Oxford University Press, Oxford (1995)

---

[3] On a technical note, NC can be defined as $\cup_{k=0}^{\infty} NC^k$, where NC^k is the class of problems solvable on a PRAM that runs for $O(\log(n))^k$ time and uses polynomial processors/space, in input length n. Equivalently NC^k can be defined as those problems solvable by circuits of $O(\log(n))^k$ depth (parallel time), and polynomial size. AC^k, and TC^k are defined analogously, except we use unbounded fan-in gates, and threshold gates, respectively.

13. Grover, L.K.: A fast quantum mechanical algorithm for database search. In: Proc. 28th Annual ACM Symposium on Theory of Computing, May 1996, pp. 212–219 (1996)
14. Karp, R.M., Ramachandran, V.: Parallel algorithms for shared memory machines. In: van Leeuwen, J. (ed.) Handbook of Theoretical Computer Science, ch. 17, vol. A, pp. 869–941. Elsevier, Amsterdam (1990)
15. Lenslet Labs. Enlight256. White paper report, Lenslet Ltd., 6 Galgalei Haplada St, Herzelia Pituach, 46733 Israel (November 2003)
16. Naughton, T., Javadpour, Z., Keating, J., Klíma, M., Rott, J.: General-purpose acousto-optic connectionist processor. Optical Engineering 38(7), 1170–1177 (1999)
17. Naughton, T.J.: Continuous-space model of computation is Turing universal. In: Bains, S., Irakliotis, L.J. (eds.) Critical Technologies for the Future of Computing, San Diego, California, August 2000. Proceedings of SPIE, vol. 4109, pp. 121–128 (2000)
18. Naughton, T.J.: A model of computation for Fourier optical processors. In: Lessard, R.A., Galstian, T. (eds.) Optics in Computing 2000, Quebec, Canada, June 2000. Proc. SPIE, vol. 4089, pp. 24–34 (2000)
19. Naughton, T.J., Woods, D.: Optical computing. In: Meyers, R.A. (ed.) Encyclopedia of Complexity and System Science. Springer, Heidelberg (to appear)
20. Naughton, T.J., Woods, D.: On the computational power of a continuous-space optical model of computation. In: Margenstern, M., Rogozhin, Y. (eds.) MCU 2001. LNCS, vol. 2055, pp. 288–299. Springer, Heidelberg (2001)
21. Parberry, I.: Parallel complexity theory. Wiley, Chichester (1987)
22. Pratt, V.R., Stockmeyer, L.J.: A characterisation of the power of vector machines. Journal of Computer and Systems Sciences 12, 198–221 (1976)
23. Reif, J.H., Tyagi, A.: Efficient parallel algorithms for optical computing with the Discrete Fourier transform (DFT) primitive. Applied Optics 36(29), 7327–7340 (1997)
24. Schönhage, A.: On the power of random access machines. In: Maurer, H.A. (ed.) ICALP 1979. LNCS, vol. 71, pp. 520–529. Springer, Heidelberg (1979)
25. Shaked, N.T., Messika, S., Dolev, S., Rosen, J.: Optical solution for bounded NP-complete problems. Applied Optics 46(5), 711–724 (2007)
26. Shaked, N.T., Simon, G., Tabib, T., Mesika, S., Dolev, S., Rosen, J.: Optical processor for solving the traveling salesman problem (TSP). In: Javidi, B., Psaltis, D., Caulfield, H.J. (eds.) Proc. of SPIE, Optical Information Systems IV, August 2006, vol. 63110G (2006)
27. Shaked, N.T., Tabib, T., Simon, G., Messika, S., Dolev, S., Rosen, J.: Optical binary-matrix synthesis for solving bounded NP-complete combinatorical problems. Optical Engineering 46(10), 108201–1–108201–11 (2007)
28. van Emde Boas, P.: Machine models and simulations. In: van Leeuwen, J. (ed.) Handbook of Theoretical Computer Science, ch. 1, vol. A, Elsevier, Amsterdam (1990)
29. Van der Lugt, A.: Optical Signal Processing. Wiley, New York (1992)
30. Woods, D.: Computational complexity of an optical model of computation. PhD thesis, National University of Ireland, Maynooth (2005)
31. Woods, D.: Upper bounds on the computational power of an optical model of computation. In: Deng, X., Du, D. (eds.) ISAAC 2005. LNCS, vol. 3827, pp. 777–788. Springer, Heidelberg (2005)
32. Woods, D.: Optical computing and computational complexity. In: Calude, C.S., Dinneen, M.J., Păun, G., Rozenberg, G., Stepney, S. (eds.) UC 2006. LNCS, vol. 4135, pp. 27–40. Springer, Heidelberg (2006)

33. Woods, D., Gibson, J.P.: Complexity of continuous space machine operations. In: Cooper, S.B., Löwe, B., Torenvliet, L. (eds.) CiE 2005. LNCS, vol. 3526, pp. 540–551. Springer, Heidelberg (2005)
34. Woods, D., Gibson, J.P.: Lower bounds on the computational power of an optical model of computation. In: Calude, C.S., Dinneen, M.J., Păun, G., Pérez-Jímenez, M.J., Rozenberg, G. (eds.) UC 2005. LNCS, vol. 3699, pp. 237–250. Springer, Heidelberg (2005)
35. Woods, D., Naughton, T.J.: An optical model of computation. Theoretical Computer Science 334(1-3), 227–258 (2005)

The Use of Hilbert-Schmidt Decomposition for Implementing Quantum Gates

Y. Ben-Aryeh

Physics Department, Technion, Haifa 32000, Israel
phr65yb@physics.technion.ac.il

Abstract. It is shown how to realize quantum gates by decomposing the gates into summation of unitary matrices where each of these matrices is given by a tensor multiplication of the unit and Pauli 2x2 spin matrices. It is assumed that each of these matrices is operating on a different copy of the quantum states produced by 'quantum encoders' with a certain probability of success. The use of the present probabilistic linear optics' method for realizing quantum gates is demonstrated by the full analysis given for the control phase shift gate, but the use of the present method for other gates, including the control-not gate, is also discussed.

1 Introduction

A quantum bit (qubit) is a two-level quantum system described by a two-dimensional complex Hilbert space [1,2]. The computational qubit state is described by a superposition of normalized and orthogonal states of a two-level quantum system denoted as |0> and |1>. In the present study photonic qubits are used where |0> and |1> represent horizontal |H>and vertical |V> polarized photons, respectively. In order to implement general quantum computational processes one needs to apply control operations. In the present work we are interested in the implementation of quantum gates with two input qubits, known as the control qubit and the target qubit, respectively. The control qubit (A) is not changed by the quantum gate, but a certain linear unitary transformation is performed on the target qubit (B) if and only if the control bit is set to |1>. Optics seems to be a good candidate for achieving two-qubit quantum gates. Unfortunately, such gates are quite difficult to implement experimentally since the state of the control qubit should affect the second target qubit and this requires strong interactions between single photons. Such interactions need high nonlinearities well beyond what is available experimentally.

Recently it has been shown by Knill, Lafllamme and Milburn [3] that probabilistic quantum logic operations can be performed using only linear optical elements, additional photons (ancilla), and post selection based on the single photon detectors. This idea has been implemented in various studies [4] and in particular Pittman, Jacobs and Franson [5-7] constructed a variety of quantum logic gates by using polarizing beam splitters (PBS) that completely transmit one state of polarization and totally reflect the orthogonal state of polarization. These methods overcome the complications introduced by using non-linear optics for realizing quantum gates, but on the other hand their nature is probabilistic throwing away a part of the measurements.

S. Dolev, T. Haist, and M. Oltean (Eds.): OSC 2008, LNCS 5172, pp. 87–97, 2008.

Probabilistic 'quantum encoding' processes have been realized experimentally and used for designing various quantum gates transformations [5-7]. The encoder consists primarily of a polarizing beam splitter (PBS) and resource pair of entangled photons in the Bell state $\left|\phi^{+}\right\rangle=\left(1/\sqrt{2}\right)\left(\left|00\right\rangle+\left|11\right\rangle\right)$ [7]. For the quantum encoder the input qubit of a single photon, in a general polarization state $\alpha\left|0\right\rangle+\beta\left|1\right\rangle$, and one member of the entangled resource pair are mixed at the PBS oriented in the HV basis. There are three output ports of the quantum encoder, including two output ports for the PBS and one output port for the second member of the entangled resource pair. Detection of one photon by 'gating detector' in one output port of the PBS signals the fact that the two remaining photons are exiting the device in the other two output ports. Because the PBS transmits $H-$polarized photons and reflects $V-$polarized photons it can be shown [5,7] that the output state is of the form

$$\left|\psi\right\rangle_{out}=\frac{1}{\sqrt{2}}\left(\alpha\left|000\right\rangle+\beta\left|111\right\rangle\right)+\frac{1}{\sqrt{2}}\left|\psi_{\perp}\right\rangle, \tag{1}$$

where $\left|\psi_{\perp}\right\rangle$ represents combinations of states orthogonal to the condition of finding one and only one (1AO1) in the gating detector. In order to implement the quantum encoding process we accept the remaining outputs only when the condition 1AO1 is satisfied. In order to have only the condition 1AO1 and erase any additional information obtained by the gating detector the encoding is completed by accepting the output only when the gating detector measures exactly one photon in a polarization basis rotated by 45^{0} from the HV basis [5]. Under these circumstances and ideal conditions, which occur with probability of 1/2 , the device realizes the encoding [7]:

$$\alpha\left|0\right\rangle+\beta\left|1\right\rangle\rightarrow\alpha\left|0\right\rangle_{1}\left|0\right\rangle_{2}+\beta\left|1\right\rangle_{1}\left|1\right\rangle_{2}. \tag{2}$$

The subscripts 1 and 2 indicate different copies of each state where the copied wavepackets are located in different places. Under ideal conditions the probability of success of the encoding process is 1/2. The encoding device is described in Fig. 1 of [7]. One should notice that the encoding transformation (2) obtained by post selection is different from the 'cloning' transformation [8]

$$\alpha\left|0\right\rangle+\beta\left|1\right\rangle\rightarrow\left(\alpha\left|0\right\rangle+\beta\left|1\right\rangle\right)_{1}\left(\alpha\left|0\right\rangle+\beta\left|1\right\rangle\right)_{2}$$

which is not allowed.

One should notice that the state $\dfrac{\left|0\right\rangle_{1}\left|0\right\rangle_{2}+\left|1\right\rangle_{1}\left|1\right\rangle_{2}}{\sqrt{2}}$ is the well known entangled state which includes certain quantum correlations ,i.e, if the first photon is in the state |0> (horizontal polarized photon) then the second photon is also in the state |0> while if the first photon is in the state |1> (vertical polarized photon) then the second photon is also in the state |1>. As is well known quantum entanglement is a fundamental resource for quantum computation processes [1,2]. The encoding transformation (2) produces a general entangled state and we would like to exploit such entangled states for implementing quantum gates.

A new method is developed in the present work for implementing quantum gates, based on the use of quantum encoders, which is basicly different from that presented in the previous works [5-7]. As is well known any unitary matrix can be decomposed into summation of tensor products of Pauli and unit spin matrices [2]. The application of such decomposition for the realization of quantum gates is quite problematic since in quantum computation we should use multiplications of unitary operators and not summations of them. However, there is a certain trick by which such decomposition can implement quantum gates. By using quantum encoders [5-7] we can 'copy' each state in the qubit superposition. Then each matrix in the above decomposition of the unitary gate operates on a different copy and by *adding* the results in the different copies we implement the corresponding gate. It should be apparent that the quantum encoders which are based on probabilistic detection procedures [6,7] and which have been realized experimentally are *different* from 'cloning' of the qubits which is prohibited by the quantum 'no-cloning' theorem [8]. The present new method is analyzed explicitly for the $CPHASE(\theta)$ gate but the options of using it for other quantum gates are also discussed.

The present paper is arranged as follows: In Section 2 we analyze the decomposition of two-qubit gates into summation of tensor products of Pauli and unit spin matrices. In Section 3 we analyze the use of the present method for implementing the $CPHASE(\theta)$ gate. In Section 4 we discuss and summarize our results and conclusions.

2 Two Qubit Gates Described by Tensor Products of Pauli and Unit Spin Matrices

For using matrix representations of quantum gates the qubits $|0\rangle$ and $|1\rangle$ are described by the following column vectors

$$|0\rangle \equiv \begin{pmatrix} 1 \\ 0 \end{pmatrix} \quad ; \quad |1\rangle \equiv \begin{pmatrix} 0 \\ 1 \end{pmatrix}. \tag{3}$$

The linear transformations operating on these single-qubit column vectors are given by multiplying them by unitary matrices of dimension 2x2. These matrices can be represented by linear combination of the four spin matrices:

$$I = \begin{pmatrix} 1 & 0 \\ 0 & 1 \end{pmatrix} \quad , \sigma_1 = \begin{pmatrix} 0 & 1 \\ 1 & 0 \end{pmatrix} \quad , \sigma_2 = \begin{pmatrix} 0 & -i \\ i & 0 \end{pmatrix} \quad , \sigma_3 = \begin{pmatrix} 1 & 0 \\ 0 & -1 \end{pmatrix} \tag{4}$$

where I is the two-dimensional unit matrix, and $\sigma_1, \sigma_2, \sigma_3$ are the Pauli spin matrices.

The two-qubit state can be given by four dimensional column vectors

$$|00\rangle \equiv \begin{pmatrix} 1 \\ 0 \end{pmatrix} \otimes \begin{pmatrix} 1 \\ 0 \end{pmatrix} \equiv \begin{pmatrix} 1 \\ 0 \\ 0 \\ 0 \end{pmatrix} \quad ; \quad |01\rangle \equiv \begin{pmatrix} 1 \\ 0 \end{pmatrix} \otimes \begin{pmatrix} 0 \\ 1 \end{pmatrix} \equiv \begin{pmatrix} 0 \\ 1 \\ 0 \\ 0 \end{pmatrix} ,$$

$$|10\rangle \equiv \begin{pmatrix} 0 \\ 1 \end{pmatrix} \otimes \begin{pmatrix} 1 \\ 0 \end{pmatrix} \equiv \begin{pmatrix} 0 \\ 0 \\ 1 \\ 0 \end{pmatrix} \quad ; \quad |11\rangle \equiv \begin{pmatrix} 0 \\ 1 \end{pmatrix} \otimes \begin{pmatrix} 0 \\ 1 \end{pmatrix} \equiv \begin{pmatrix} 0 \\ 0 \\ 0 \\ 1 \end{pmatrix} \tag{5}$$

where in the ket states on the left handside of these equations the first and second number denote the state of the first and second qubit, respectively. The sign $\otimes$ represents tensor product where the two-qubit states can be described by tensor products of the first and second qubit column vectors.

CNOT gate is given by

$$CNOT|x,y\rangle = |x, x \oplus y\rangle \quad , \quad (x = 0,1 \, ; \, y = 0,1). \tag{6}$$

and $\otimes$ indicates addition modulo 2 . This gate flips the state of the target qubit y if the control qubit x is in the state $|1\rangle$ and does nothing if the control qubit is in the state $|0\rangle$. *CNOT* can be represented by a unitary 4x4 dimensional matrix operating on the above four dimensional vectors :

$$CNOT = \begin{pmatrix} 1 & 0 & 0 & 0 \\ 0 & 1 & 0 & 0 \\ 0 & 0 & 0 & 1 \\ 0 & 0 & 1 & 0 \end{pmatrix} . \tag{7}$$

The 4x4 unitary matrix *CPHASE*(θ) gate is given by

$$CPHASE(\theta) = \begin{pmatrix} 1 & 0 & 0 & 0 \\ 0 & 1 & 0 & 0 \\ 0 & 0 & 1 & 0 \\ 0 & 0 & 0 & e^{i\theta} \end{pmatrix} \tag{8}$$

This gate (which does not have a classical analog [1]) applies a phase shift for the state $|1,1\rangle$ giving

$$CPHASE(\theta)|1,1\rangle = e^{i\theta}|1,1\rangle, \tag{9}$$

and does nothing if it operates on other states.

The CNOT gate is a standard component in computational circuits analysis [1,2]. Quantum computational circuits in which the *control phase shift gate*-$CPHASE(\theta)$ is inserted as one component has been analyzed in the literature (see e.g. [2], Figures (3.5) and (3.6) on page 116).

Any two-qubit gate can be expressed in the Hilbert-Schmidt (*HS*) representation [9] as

$$U_2 = \sum_{j,k=0}^{3} t_{j,k} \sigma_j \otimes \sigma_k \,, \tag{10}$$

where by taking into account the properties of the spin matrices we find

$$t_{l,m} = \frac{1}{4} Tr\left[U_2 \bullet (\sigma_l \otimes \sigma_m)\right] \tag{11}$$

The point $\bullet$ represents the ordinary matrix multiplication, the sign $\otimes$ denotes tensor product and the notation Tr represents the trace operation. $4t_{l,m}$ is given by the trace of the ordinary matrix multiplication of the four-dimensional matrix U_2 by the four dimensional matrix $(\sigma_l \otimes \sigma_m)$. In deriving (11) we use the relations

$$Tr\left[(\sigma_j \otimes \sigma_k) \bullet (\sigma_l \otimes \sigma_m)\right] = 4\delta_{j,l}\delta_{k,m} \tag{12}$$

While for a general two-qubit gate 16 elements of $t_{j,k}$ might be different from zero, for the main basic two-qubit gates only 4 elements $t_{j,k}$ are different from zero. We find by straightforward calculations for the CNOT unitary matrix of (7):

$$CNOT = \frac{1}{2}\left[\sigma_3 \otimes (I - \sigma_1)\right] + \frac{1}{2}\left[I \otimes (\sigma_1 + I)\right]. \tag{13}$$

For the $CPHASE(\theta)$ gate of (8) we get:

$$CPHASE(\theta) = \kappa(I \otimes I) + \lambda(\sigma_3 \otimes \sigma_3) + \mu(I \otimes \sigma_3) + \nu(\sigma_3 \otimes I) \tag{14}$$

where

$$\lambda = \frac{1}{4}(e^{i\theta} - 1) \quad , \quad \mu = \nu = -\lambda \quad , \quad \kappa = 1 + \lambda \quad . \tag{15}$$

Each of Eqs. (13,14) includes four tensor products where the first and second 2x2 matrix in each tensor product operates on the first and second qubit column vectors, respectively. For each qubit we can use the relations:

$$\sigma_3|0\rangle = |0\rangle \;\; ; \;\; \sigma_3|1\rangle \equiv -|1\rangle \;\; ; \;\; \sigma_1|0\rangle = |1\rangle \;\; ; \;\; \sigma_1|1\rangle = |0\rangle \;\; ;$$
$$\sigma_2|0\rangle = i|1\rangle \;\; ; \;\; \sigma_2|1\rangle = -i|0\rangle \;\; ; \;\; I|0\rangle = |0\rangle \;\; ; \;\; I|1\rangle = |1\rangle \tag{16}$$

In polarization optics the states represented by the column vectors of (3) are known as Jones vectors [10]. By using Jones calculus it is quite easy to implement the unitary transformations of (16) operating on single qubit column vectors.(See polarization optics transformations obtained by Jones calculus, including the effects of *half- and quarter-wave retardation plates* , and the general 2x2 unitary matrices transformation (1.5-11) of [10]).

As explained in the introduction the application of the decompositions (13,14) for the realization of quantum gates is quite problematic but they can implement quantum gates by the use of quantum encoders, as described in the following analysis.

3 Realization of $CPHASE(\theta)$ Gate by Quantum Encoders

An input two-qubit state can be written as

$$|\psi\rangle_{in} = \{\alpha|0\rangle_A + \beta|1\rangle_A\}\{\gamma|0\rangle_B + \delta|1\rangle_B\}, \tag{17}$$

where the subscripts A and B refer to two separated qubits. The complex amplitudes for the first and second qubit are given by α and β , and γ and δ, respectively. The $CPHASE(\theta)$ is defined as leading to the output state

$$|\psi\rangle_{out} = \alpha\gamma|0\rangle_A|0\rangle_B + \alpha\delta|0\rangle_A|1\rangle_B + \beta\gamma|1\rangle_A|0\rangle_B + \beta\delta e^{i\theta}|1\rangle_A|1\rangle_B \tag{18}$$

The first qubit (A) acts as a control and its value is unchanged on the output. In case that the first control qubit is in the |0> state nothing happens to the second target qubit (B). In case that the first control qubit is in the |1> additional   phase θ is inserted between the |0> state and the |1> state of the second target qubit, and we *define* this additional phase to be inserted in the |1) state (but take into account that only the relative phase is important). In quantum computational circuits the case $CMINUS = CPHASE(\pi)$ is especially important [2]. In the following analysis it is shown how to implement the transformation (18) by using quantum encoders.

By using quantum encoders [7], as explained in the introduction, we can copy two times each input state transforming (17) into

$$|\psi\rangle_{in} = \left[\alpha\{|0\rangle_{A1}|0\rangle_{A2}\} + \beta\{|1\rangle_{A1}|1\rangle_{A2}\}\right] \times \left[\gamma\{|0\rangle_{B1}|0\rangle_{B2}\} + \delta\{|1\rangle_{B1}|1\rangle_{B2}\}\right] \tag{19}$$

In (19) we get multiplications of four states since each of the two-states denoted by the subscripts A and B has been copied twice by the quantum encoders and these copies are indicated by adding the subscripts one and two.

The input state $|\psi\rangle_{in}$ of (19) can be rearranged as

$$
\begin{aligned}
|\psi\rangle_{in} &= \alpha\gamma\{|0\rangle_{A1}|0\rangle_{B1}\}\{|0\rangle_{A2}|0\rangle_{B2}\} + \alpha\delta\{|0\rangle_{A1}|1\rangle_{B1}\}\{|0\rangle_{A2}|1\rangle_{B2}\} \\
&+ \beta\gamma\{|1\rangle_{A1}|0\rangle_{B1}\}\{|1\rangle_{A2}|0\rangle_{B2}\} + \beta\delta\{|1\rangle_{A1}|1\rangle_{B1}\}\{|1\rangle_{A2}|1\rangle_{B2}\}
\end{aligned}
\tag{20}
$$

For each four-state multiplication of (20) the two-states given in the first curled bracket which are indicated by the subscripts A_1 and B_1 are copied into equivalent two-states given in the second curled bracket which are indicated by the subscripts A_2 and B_2.

Eq. (14) can also be written as a summation of two unitary matrices

$$
CPHASE(\theta) = \left(I \otimes (\kappa I + \mu\sigma_3)\right) + \left(\sigma_3 \otimes (\lambda\sigma_3 + \nu I)\right)
\tag{21}
$$

By using the decomposition of (21) into the summation of two 4X4 unitary matrices and the relations (16), we will assume that the unitary matrix $\left(I \otimes (\kappa I + \mu\sigma_3)\right)$ will operate on the two-states given in the first curled brackets of (20) with subscripts A_1 and B_1 leading to the transformations:

$$
\{|0\rangle_{A1}|0\rangle_{B1}\} \rightarrow |0\rangle_{A1}\{(\kappa+\mu)|0\rangle_{B1}\}; \quad \{|0\rangle_{A1}|1\rangle_{B1}\} \rightarrow |0\rangle_{A1}\{(\kappa-\mu)|1\rangle_{B1}\}
$$
$$
\{|1\rangle_{A1}|0\rangle_{B1}\} \rightarrow |1\rangle_{A1}(\kappa+\mu)|0\rangle_{B1}; \quad \{|1\rangle_{A1}|1\rangle_{B1}\} \rightarrow |1\rangle_{A1}\{(\kappa-\mu)|1\rangle_{B1}\}
\tag{22}
$$

and that the unitary matrix $\left(\sigma_3 \otimes (\lambda\sigma_3 + \nu I)\right)$ will operate on the two-states given in the second curled brackets of (20) with subscripts A_2 and B_2 leading to the transformations:

$$
\{|0\rangle_{A2}|0\rangle_{B2}\} \rightarrow |0\rangle_{A2}\{(\lambda+\nu)|0\rangle_{B2}\}; \quad \{|0\rangle_{A2}|1\rangle_{B2}\} \rightarrow |0\rangle_{A2}\{(\nu-\lambda)|1\rangle_{B2}\}
$$
$$
\{|1\rangle_{A2}|0\rangle_{B2}\} \rightarrow -|1\rangle_{A2}\{(\lambda+\nu)|0\rangle_{B2}\}; \quad \{|1\rangle_{A2}|1\rangle_{B2}\} \rightarrow |1\rangle_{A2}\{(\lambda-\nu)|1\rangle_{B2}\}
\tag{23}
$$

Such processes can be implemented experimentally due to different locations of the two-states so that the operation of the unitary matrix $CPHASE(\theta)$ has been decomposed here into the summation of two unitary processes each operating on a different copy of the two-states.

Performing the transformations (22-23) on the input state (20) we get :

$$|\psi\rangle_{out} = \alpha\gamma\left\{|0\rangle_{A1}\left[(\kappa+\mu)|0\rangle_{B1}\right]\right\}\left\{|0\rangle_{A2}\left[(\lambda+\nu)|0\rangle_{B2}\right]\right\}$$
$$+\alpha\delta\left\{|0\rangle_{A1}\left[(\kappa-\mu)|1\rangle_{B1}\right]\right\}\left\{|0\rangle_{A2}\left[(\nu-\lambda)|1\rangle_{B2}\right]\right\}$$
$$+\beta\gamma\left\{|1\rangle_{A1}\left[(\kappa+\mu)|0\rangle_{B1}\right]\right\}\left\{|1\rangle_{A2}\left[-(\lambda+\nu)|0\rangle_{B2}\right]\right\}$$
$$+\beta\delta\left\{|1\rangle_{A1}\left[(\kappa-\mu)|1\rangle_{B1}\right]\right\}\left\{|1\rangle_{A2}\left[(\lambda-\nu)|1\rangle_{B2}\right]\right\}$$

$$(24)$$

Here the phases of the states with subscripts A_1 and A_2 are assumed to be positive and relative to them the phases of the qubits with subscribts B_1 and B_2 are given. One should take into account that by the copying procedure the input and correspondingly the output states were doubled.

We can consider (24) as a certain implementation of the $CPHASE(\theta)$ gate where the control operation of this gate has been decomposed into two equal control qubits. We find that the states with subscript A_1 are equal to those with subscript A_2, both can be considered as equal to the control qubit which is not changed by the quantum gate. The target states have been decomposed here into two *different* target states denoted by the subscripts B_1 and B_2. We get a *relative* phase of the target state denoted by subscript B_1 relative to the control state denoted by subscript A_1, and we get a relative phase of the target state denoted by subscript B_2 relative to the control state denoted by subscript A_2. When we add these two relative phases, which can be obtained in two separated experiments, the $CPHASE(\theta)$ gate is realized as described by the following correspondences:

$$|0\rangle_{A1}|0\rangle_{A2} \rightarrow 2|0\rangle_A \quad ; \quad |1\rangle_{A1}|1\rangle_{A2} \rightarrow 2|1\rangle_A \quad ;$$
$$\left[(\kappa+\mu)|0\rangle_{B1}\right]\left[(\lambda+\nu)|0\rangle_{B2}\right] \rightarrow (\kappa+\mu+\lambda+\nu)|0\rangle_B \quad ;$$
$$\left[(\kappa-\mu)|1\rangle_{B1}\right]\left[(\nu-\lambda)|0\rangle_{B2}\right] \rightarrow (\kappa-\mu+\nu-\lambda)|0\rangle_B \quad ; \quad (25)$$
$$\left[(\kappa+\mu)|0\rangle_{B1}\right]\left[-(\nu+\lambda)|0\rangle_{B2}\right] \rightarrow (\kappa+\mu-\nu-\lambda)|0\rangle_B \quad ;$$
$$\left[(\kappa-\mu)|1\rangle_{B1}\right]\left[(\lambda-\nu)|1\rangle_{B2}\right] \rightarrow (\kappa-\mu+\lambda-\nu)|0\rangle_B$$

Using the relations (15) we get

$$(\kappa+\mu+\lambda+\nu)=1 \quad ; \quad (\kappa-\mu+\nu-\lambda)=1 \quad ;$$
$$(\kappa+\mu-\nu-\lambda)=1 \quad ; \quad (\kappa-\mu+\lambda-\nu)=1+4\lambda=e^{i\theta}$$

$$(26)$$

Substituting (25,26) into (24) we get

$$\frac{|\psi\rangle_{out}}{2} = \alpha\gamma|0\rangle_A|0\rangle_B + \alpha\delta|0\rangle_A|1\rangle_B + \beta\gamma|1\rangle_A|0\rangle_B + \beta\delta e^{i\theta}|1\rangle_A|1\rangle_B \quad (27)$$

which is equivalent to the transformation given by (18) (up to the unimportant factor 2). The transformation in the doubled space of (20) to the output (24) with the above correspondences leads in a certain special way to implementation of the $CPHASE(\theta)$ gate.

One should take into account that $CPHASE(\theta)$ has been implemented by one to one correspondence of (20) to (24) so that such implementation is mainly in 'principle'. One might also perform the addition of the relative phases in interference experiments leading to relations (26) and then the implementation will be also in 'practice'. The relations (25,26) are given by the addition of the amplitudes like those given by interference experiments and are basicly different from the addition of logic states numbers [2]. The transformation in the doubled space of the input state $|\psi\rangle_{in}$ of (19) to the output state $|\psi\rangle_{out}$ of (24) realizes the quantum gate since we can transform back the doubled output state to the ordinary $CPHASE(\theta)$ output two-qubit state.

4 Summary, Discussion and Conclusion

The use of probabilistic logic operations has been developed for implementing quantum gates. It has been shown that quantum gates can be decomposed into summation of tensor products of unit and Pauli 2x2 spin matrices. Such HS decomposition has been applied for the $CPHASE(\theta)$ gate leading to a summation of two unitary matrices of dimension 4x4. In the present method each 4x4 matrix operates on a different copy of the two-states produced by quantum encoders. By adding the relative phases in the two two-states' copies the $CPHASE(\theta)$ gate is realized.

The same technique that has been described in the present work for implementing the $CPHASE(\theta)$ gate can be used also for implementing the $CNOT$ gate. The encoding process for the quantum states is the same for the two cases. The only difference is that the decomposition of the $CPHASE(\theta)$ gate by (14) is replaced by the decomposition of the $CNOT$ by (13). Replacing the operations of the unitary matrices of (14) on the quantum states, by those of (13) one can use a similar procedure for implementing the $CNOT$ gate. While the $CPHASE(\theta)$ gate is realized by adding the relative phases of the two target states, the $CNOT$ gate can be realized by adding the polarization states of the two target states. Thus, we have shown therefore a new method for implementing both $CNOT$ and $CPHASE(\theta)$ gates [1,2].

In the present work we have used the HS decomposition by which the quantum gate is decomposed into a summation of unitary matrices where each of these matrices is given by tensor products of Pauli and unit spin matrices. It has been shown that by operating with each of these matrices on a different copy of the states in each superposition and by adding the results for the different copies the quantum gate is realized. The analyses for $CPHASE(\theta)$ and $CNOT$ gates are relatively simple due to the fact that for these gates the decomposition can include only two such matrices. For other gates the HS decomposition might include the summation of more unitary matrices so that the corresponding copying processes by quantum encoders should be more complicated. However, the present analysis becomes quite general if we consider the fact that any quantum gate can be obtained by the combinations of single-qubit gates and two-qubit $CPHASE(\theta)$ and $CNOT$ gates [1,2].

Quantum encoding processes which are obtained by using probabilistic transformations have been already applied successfully in the experiments reported in [5-7]. In the present work the use of quantum encoders has been developed for implementing quantum gates by new methods using the HS decomposition which are different from those used previously and these methods should therefore be of interest both theoretically and experimentally.

Any quantum computational process is described by a certain circuit assuming any initial input state and its end to be measured. The quantum circuits described in the present analysis seem to be different from the conventional ones. However, the initial state assumed in our analysis and its end to be measured are equivalent to the corresponding conventional two-qubit gates. Therefore we find that the present method has developed certain realizations of the quantum gates.

References

1. Nielsen, M.A., Chuang, I.L.: Quantum Computation and Quantum Information. Cambridge University Press, Cambridge (2001)
2. Benenti, G., Casati, G., Strini, G.: Principles of Quantum Computation and Information, Basic concepts, vol. 1. World Scientific, Singapore (2005)
3. Knill, E., Laflamme, R., Milburn, G.J.: A scheme for efficient quantum computation with linear optics. Nature (London) 409, 46–52 (2001)
4. Kok, P., Munro, W.J., Nemoto, K., Ralph, T.C., Dowling, J.P., Milburn, G.J.: Linear optical quantum computing with photonic qubits. Rev. Mod. Phys. 79, 135–174 (2007)
5. Pittman, T.B., Jacobs, B.C., Franson, J.D.: Probabilistic quantum logic operations using polarization beam splitters. Phys. Rev. A 64(062311), 1–9 (2001)
6. Pittman, T.B., Jacobs, B.C., Franson, J.D.: Demonstration of non-deterministic quantum logic operations using linear optical elements. Phys. Rev. Lett. 88(257902), 1–4 (2002)
7. Pittman, T.B., Jacobs, B.C., Franson, J.D.: Probabilistic quantum encoder for single-photon qubits. Phys. Rev. A 69(042306), 1–4 (2004)

8. Peres, A.: Quantum Theory: Concepts and Methods. Kluwer, London (1998)
9. Ben-Aryeh, Y., Mann, A., Sanders, B.C.: Empirical state determination of entangled two-level systems and its relation to information theory. Foundations of Physics 29, 1963–1975 (1999)
10. Yariv, A.: Optical Electronics. Saunders College Publishing, NewYork (1991)

A Method for Modulo Operation by Use of Spatial Parallelism

Kouichi Nitta, Nobuto Katsuta, and Osamu Matoba

Department of Computer Science and Systems Engineering,
Graduate School of Engineering, Kobe University
Rokkodai 1-1, Nada, Kobe 657-8501, Japan
nitta@kobe-u.ac.jp, katsuta@brian.cs.kobe-u.ac.jp,
matoba@kobe-u.ac.jp

Abstract. An optical method for modulo operations has been proposed. This method utilizes phase modulation of light wave. This method can be applied to modulo multiplication which is an important operation in an algorithm for prime factorization. Optical parallel processing based on the method is implemented with a Michelson interferometer. This report shows that this method is effective in prime factorization. Especially, we study on suitability between large scale data processing for the prime factorization and the proposed method.

Keywords: Parallel processing, phase modulation, optical interference, modulo operation, prime factorization, spatial parallelism.

1 Introduction

As is well known, optical signals have various advantaged features for information processing. Broad bandwidth and huge capability for data storage are mentioned as examples of such features. Also, spatial parallelism is one of the promising characteristics in optical information processing.

Recently, some optical methods for problems requiring exponential computational costs with electronic processing have been proposed. In Ref. [1], a method for the Hamiltonian path problem is reported. This method utilizes delay of the rays. Also, two solutions for the traveling salesman problem have been developed. One is based on white light interferometry with fiber optics [2]. In the other methods, a set of network is represented as a binary matrix and the output is obtained by a matrix vector multiplication [3]. The multiplication process is realized with a joint transform correlator.

In such a situation, we have proposed an optical method for parallel modulo operations [4]. One of the advantaged features of the proposed method is based on spatial parallelism of light. This method gives wave fields corresponding to results of modulo operation by modulating phase of light wave. Moreover, this method is applied to modulo multiplication. Massive data processing for modulo multiplication is important in an algorithm for prime factorization [5]. The proposed method has been verified to be useful for prime factorization.

S. Dolev, T. Haist, and M. Oltean (Eds.): OSC 2008, LNCS 5172, pp. 98–103, 2008.

In this report, the principle of the method is described. And, some advantaged features of the system are discussed.

2 Modulo Operation with Optical Phase Modulation

In the factoring algorithm reported in Ref. [5], two prime numbers of a target integer are obtained with the period of modulo exponentiation as described in Eq. (1).

$$f(x) = a^x \bmod N \tag{1}$$

In this equation, N shows a target integer ($N=pq$). And a is an integer selected in pre-processing. Here, a should be satisfied with inequality (2) and Eq. (3), respectively.

$$1 < a < N \tag{2}$$

$$\gcd(a, N) = 1 \tag{3}$$

In Eq. (3), gcd (a, N) indicates the greatest common devisor between a and N. In case that the period of $f(x)$ is an odd number, p and q are given by the following two equations.

$$p = \gcd(a^{r/2} - 1, N) \tag{4}$$

$$q = \gcd(a^{r/2} + 1, N) \tag{5}$$

In the algorithm, $f(x)$ is derived in accordance with the Shonhage-Strassen algorithm [7]. Note that modulo exponentiation is derived with sequence of modulo multiplication represented as Eq. (6).

$$g(x) = yx \bmod N \tag{6}$$

Let us consider a sinusoidal wave defined as Eq. (7).

$$U(\phi) = \cos(2\pi\phi) \tag{7}$$

In Eq. (7), by setting $\varphi = yx/N$, Eq. (7) is modified as shown in Eq. (8).

$$\begin{aligned}
U(\phi) &= \cos\left(2\pi \frac{yx}{N}\right) \\
&= \cos\left(2\pi \frac{kN + g(x)}{N}\right) \\
&= \cos\left(\frac{2\pi}{N} g(x)\right)
\end{aligned} \tag{8}$$

Eq. (3) shows that wave fields corresponding to remainder are obtained by simple phase modulation.

3 Optical Hardware and Improved Solutions

3.1 Basic Architecture for Optical Implementation

An optical system for parallel processing based on the above scheme can be constructed with a Michelson interferometer as described in Fig. 1. In the system, the mirror put at one optical arm is tilted to generate desired interference signals. A tilt angle to execute parallel processing shown in Eq. (7) is given by Eq. (8).

$$\theta = \frac{1}{2}\sin^{-1}\left(\frac{y\lambda}{DN}\right)$$

(8)

In this equation, λ and D show wavelength of the light source and pixel pitch of PD array, respectively. Interference signals are observed with photodetector array. Optical path difference between pixels is ($y\lambda/N$). In accordance with procedure reported in Ref. [4], prime factorization is executed with the optical system and post processing. We have developed and demonstrated an optical system based on the architecture described in Fig. 1.

3.2 Previous Works

In the first prototype reported in Ref. [4], 640 points of $g(x)$ can be achieved in parallel. The performance of parallel processing directly depends on the array size of detectors. It has been shown that proposed method is able to give correct period of $f(x)$ with post processing even though noise signals are included in measured interference signals.

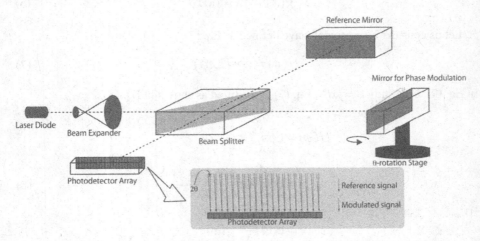

Fig. 1. Schematic diagram of an optical parallel processor for modulo operations

A method for two-dimensional parallel processing has been reported as improvement of the proposed method [6]. In the two-dimensional parallel system, both mirrors in the interferometer are controlled. One is rotated at θ-direction. And, the other is turned at α-direction. Interference patterns are generated and measured with two dimensional array of photo sensors. Note that this architecture is suitable for an area sensor. Fig. 2 shows a photograph of the constructed system. This system can achieve 1344x1024 points of parallel operations. It is shown that two dimensional processing is useful to improve processing performance dramatically.

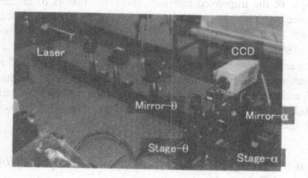

Fig. 2. Photograph of the experimental system for two dimensional parallel processing

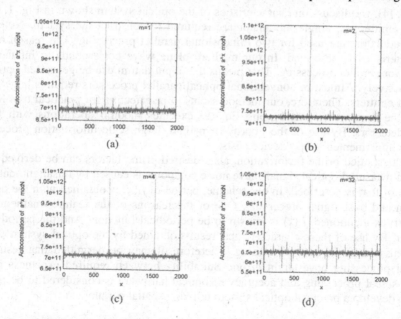

Fig. 3. Numerical analysis for auto correlation of modulo exponentiation obtained by the improved method with Eq. (9). (a)~(d) show those in case of $m=1, 2, 4, 32$, respectively.

Another method has been proposed. This method improves processing performance without change of device. In the method θ is set as described in Eq. (9).

$$\theta = \frac{1}{2}\sin^{-1}\left(\frac{m}{D}\frac{a\lambda}{N}\right)$$ (9)

Here, m must be a natural number. Therefore, this method can achieve m times of parallel processing in comparison with the first system described in Fig. 1. Fig. 3 shows an example of the improved method. There are results of numerical analysis. Horizontal and vertical axes show x and auto correlation of $f(x)$, respectively. In this case, N and a are 1643 and 300 respectively. From the graphs, it is confirmed that correct period (=780) is derived in case of $m \leqq 32$.

On the other hands, noise robustness of an optical system described in Fig. 1 is estimated [9]. In an interferometer, misalignment and noise components are unavoidable. Therefore, error tolerance of the optical system is important for estimation of the proposed method. As results of estimations, we show that the proposed method has high robustness against noise signals.

4 Discussion about Spatial Parallelism and Suitability of Mathematical Property

In Ref. [4], we discuss on characteristics of the optical system shown in Fig. 1. Only single emitter and single modulator are required to construct the system. Note that two modulators are used for two dimensional parallel processing. The reason of the characteristics is described. In our method, plane wave corresponds to input statements for parallel processing. That means that input datum can be generated optically and passively. Almost of conventional optical parallel processors requires huge numbers of emitters. Therefore, our method seems to be effective for practical use. Moreover, the improved method using Eq. (9) executes modulo operations with fewer photodetectors. By use of the improved method, large scale information processing can be implemented at less device costs.

In the studied prime factorization, also, desired prime factors can be derived even though measured optical signals have noise components caused by misalignment. The reason of that is described. In our scheme, period of $f(x)$ is obtained with the optical system and post signal processing. One of the reasons of the feature, mathematical property is mentioned. $f(x)$ is known to be periodical function. And the period is an integer. By use of these characteristics, results obtained by the optical system can be compensated in the post processing. Therefore, it may be permitted that results of optical processing have slight errors. Suitability between required exactness in the target signal processing and accuracy of optical hardware is considered to be important to develop a practical optical system utilizing spatial parallelism.

5 Summary

We have reported an optical method for parallel processing. This method is based on optical interference and is effective in prime factorization. Reasons of the effectiveness of the method have been discussed. In the discussion, we especially focus on the spatial parallelism of optical processing and suitability between mathematical characteristics and optical operations.

However, we have not yet developed a solution to execute prime factorization in polynomial time costs. To construct the solution is final goal of our research and a challenging issue.

References

1. Oltean, M.: Solving the Hamiltonian path problem with a light-based computer. Nat. Comput. 7, 57–70 (2008)
2. Haist, T., Osten, W.: An optical solution for the traveling salesman problem. Opt. Exp. 15, 10473–10482 (2007)
3. Shaked, N.T., Messika, S., Dolev, S., Rosen, J.: Optical solutions for bound NP-complete problems. Appl. Opt. 46, 711–724 (2007)
4. Nitta, K., Matoba, O., Yoshimura, T.: Parallel processing for multiplication modulo by means of phase modulation. Appl. Opt. 47, 611–616 (2008)
5. Shor, P.: Algorithms for quantum computation: Discrete logarithms and factoring Algorithms for quantum computation: Discrete logarithms and factoring. In: Proc. 35th Ann. Symp. on Foundations of Comput. Sci., vol. 1898, pp. 124–134 (1994)
6. Katsuta, N., Nitta, K., Matoba, O.: Parallel processor for modulo multiplication with optical interference. In: Technical Digest of The 13th Microoptics Conference, pp. 84–185 (2007)
7. Vedral, J., Barenco, A., Ekert, A.: Quantum networks for elementary arithmetic operations. Phys. Rev. A 54, 147–153 (1996)
8. Nitta, K., Katsuta, N., Matoba, O.: Study on processing performance of optical modulo operations J. Conf. Series (submitted)
9. Nitta, K., Katsuta, N., Matoba, O.: An optical interferometer for parallel modulo operation The review of laser engineering (accepted)

Learning at the Speed of Light: A New Type of Optical Neural Network

A. Steven Younger[1] and Emmett Redd[2]

[1] Jordan Valley Innovation Center, Missouri State University, Springfield, MO 65897 USA
[2] Department of Physics, Astronomy and Materials Science, Missouri State University
{SteveYounger,EmmettRedd}@MissouriState.edu

Abstract. Most, if not all, optical hardware-based neural networks are slow during the neural learning phase. This limitation has been not only a speed bottleneck, but it has contributed to the lack of wide-spread use of optical neural systems. We present a novel solution – Optical Fixed-Weight Learning Neural Networks. Standard neural networks learn new function mappings by the changing of their synaptic weights. However, the Fixed-Weight Neural Networks learn new mappings by dynamically changing recurrent neural signals. The (fixed) synaptic weights of the FWL-NN implement a learning "algorithm" which adjusts the recurrent signals toward their proper values.

Keywords: Optical Neural Networks, Optical Computing, Fixed-Weight Learning Neural Networks, Adaptive Neural Networks, Accommodative Neural Networks.

1 Introduction

Optical hardware is probably the fastest method of performing the forward-propagation phase of neural networks. An optical neural computer similar to those presented in [1, 2] should be able to perform over 10^{13} synaptic operations per second using current technology. Optical Neural squashing computations can now be performed on the sub-picoseconds time scale [3].

Most, if not all optical hardware schemes are slow during the neural learning phase. Optical learning has traditionally been done on a separate (non-optical) computer and the results stored on film, or required the use of a relatively slow (and/or expensive) spatial light modulator. This limitation has been not only a speed bottleneck, but it has contributed to the lack of wide-spread use of optical neural systems.

We present a different solution – Optical Fixed-Weight Learning Neural Networks (Optical FWL-NN). Standard neural networks learn new function mappings by the changing of their synaptic weights. However, the FWL-NNs learn new mappings by dynamically changing recurrent neural signals. The (fixed) synaptic weights of the FWL-NN implement learning "algorithm" which adjusts the recurrent signals toward their proper values. That is, instead of encoding a particular mapping, the synaptic weights of a FWL-NN encode how to learn any mapping (within a large, perhaps infinite, set of possible mappings).

S. Dolev, T. Haist, and M. Oltean (Eds.): OSC 2008, LNCS 5172, pp. 104–114, 2008.

We developed an optical hardware neural network to investigate the precision, alignment, calibration, speed, and algorithmic issues associated with Optical FWL-NNs. We report on the hardware design, generation of the synaptic weights, and initial results for some Fixed-Weight Learning tasks.

2   Optical Neural Hardware

Our optical hardware was not designed to be especially fast or to accommodate exceptionally large networks. It serves as a test apparatus for studying Optical Fixed-Weight Learning Neural Networks. Flexibility of use and (relatively) low cost were our main design criteria. With what we have learned, we are in the process of designing a fast, compact and expandable Optical Neural Network platform.

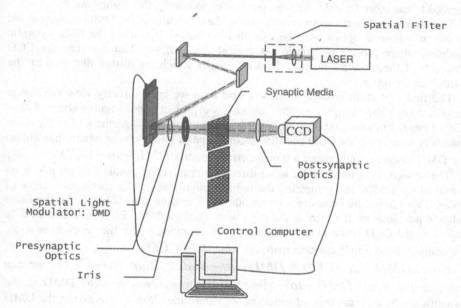

Fig. 1. Optical Neural Hardware

2.1   Hardware Overview

Figure 1 shows the optical neural hardware test apparatus. Light from a laser is expanded and directed toward a Spatial Light Modulator (SLM). The SLM creates the neural signals by modulating the intensity of a set of light beams. We used a Digital Micromirror Device (DMD) for the SLM. The DMD consists of a rectangular array of almost 1 million tiny mirrors along with drive and interfacing electronics. Under software control, each mirror can be individually set to either on (reflecting its beam toward the presynaptic optics) or off (reflecting away). The resulting signal beams pass through pre-synaptic optics and onto the synaptic medium, (35mm slide). The slide has small rectangular areas of various shades of gray that encode the synaptic weights. Synaptic multiplications are performed by the attenuations of the several light beams passing through the medium.

The attenuated optical signals are focused onto a CCD array and sent to the computer. The optical signals are spatially integrated over each region-of-interest. These dendrite signals are then sign-summed and the nonlinear squashing function applied, producing the network outputs. These last functions are currently performed in software.

2.2 Distortion, Alignment and Calibration

The 35 mm film synaptic medium can be generated with high spatial precision, as can the positions of a source neuron on the DMD and terminal neurons on the CCD. However, the pre- and post-synaptic optics generate considerable distortion. One solution would have been to create elaborate optics to eliminate these distortions. We decided to use more flexible software processing to correct the distortions.

There are two distortions to correct. First, the distortion of the DMD image caused by the pre-synaptic optics must be canceled out before it reaches the fixed synaptic medium, where precise registration is critical. Second, we must correlate the CCD positions of the (now attenuated) dendrite images which are further distorted by the post-synaptic optics.

The first distortion is hard to measure because we can't directly view the image projected onto the synaptic media. Instead, we project a rectangular array of dots (called *pegs*) from the DMD through a transparent slide and onto the CCD. By automatically measuring the CCD coordinates of the pegs, and knowing where they are on the DMD image, we computed a transformation matrix $DMD-to-CCD$.

The second distortion can be more directly measured by sending an *all pixels on* signal to the DMD, and projecting the light through a slide of a rectangular array of *holes*. These holes are clear areas on an otherwise opaque slide. They are at the same relative positions on the film as the pegs were on the DMD. By semi-automatically capturing the CCD coordinates of the holes, and knowing where they are on the slide, we computed the transformation matrix $Slide-to-CCD$.

Since $DMD-to-CCD = DMD-to-Slide \times Slide-to-CCD$, we can compute the matrix $DMD-to-Slide$, which transforms from the DMD to the synaptic medium. From this information, we determined how to *pre-distort* the DMD image in such a way that the presynaptic optics *undistorts* it and cause the DMD image to arrive properly aligned with the synaptic slide. That is, to make the pegs match the holes.

We performed the matrix calculations with up-to linear, quadratic and cubic terms in the matrices. The cubic calculation performed the best. It can correct for translations, rotations, stretching, keystoning, pincushioning and barrel distortions. We found that 42 pegs/holes (6 x 7) gave good results. This created over-determined transformation matrices. We used the Moore-Penrose pseudo inverse to find a least-squares solution.

Figure 2 shows a CCD image of 30 synapses projected from the DMD through the slide and onto the CCD. The gray levels differ due to the attenuation by the slide. Notice substantial distortion of the (originally rectangular) areas. The white boxes

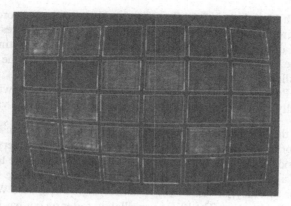

Fig. 2. Actual and Computed Regions-of-Interest

around the gray areas show where the projected synapses are expected to be -- based on the matrix calculations. Note that the computed locations are in excellent agreement with the actual image projections, although they are substantially distorted.

2.3 Encoding the Synaptic Weights

Film has a large storage capacity of 10 MegaPixels for a 35mm slide. A high-quality Holographic Plate can store an order of magnitude more information [2]. To reduce problems associated with film's very non-linear grayscale, we used a binary pixel area encoding for our synapses. A synaptic weight of $W \in [0...1]$ will have a portion of W of randomly selected pixels within its area set to clear, and $1-W$ of its pixels set to opaque. This worked out to be over 16 significant bits of gray level precision for the synapses in our networks.

The accuracy of the film medium proved to be more problematic. The lack of reproducibility of the actual gray level from slide-to-slide, and even between different areas of the same slide, was a major difficulty to be solved for fixed-weight learning to be successful.

Our solution this problem was to calibrate each dendritic area individually. Doing this *automatic gain control* once-per-phase also solved problems of the laser light source intensity and beam profile drifting over time.

2.4 Encoding the Neural Signal Intensity

The individual DMD pixels can be in one of two states – 0 or 1, on or off. This limits the type of intensity modulation schemes that can be used with this design. We tested three modulation schemes.

1. Area Pixelation (AP): Because the source areas contain many pixels (several thousand in our examples) a gray level of between zero and one can be produced by turning on the number of pixels proportional to the desired signal intensity. These pixels were uniformly and stochastically selected over each source rectangle.

2. Pulse Width Modulation (PWM) – generate N time slices according to the number of required significant bits of the signal. (e.g. 256 time slices for 8 significant bits.) Turn all the pixels in a rectangular area on a number of time slices proportional to the desired signal value, and then turn them off for the remaining pulse train time slices.

3. Stochastic Pulse (SP) – same as PWM, except the pixels are switched on and off stochastically in the proper portions to create the desired signal value [10].

The AP had the advantage of being much faster because it only used one time slice instead of 256 or more. However, the response of the system was very non-linear.

The PWM and SP methods both have the advantage of being able to adjust the number of required signal significant bits by changing the number of time slices. Speed can be sacrificed for accuracy, or vice versa. The accuracy of both the PWM and SP methods was essentially the same. The PWM was slightly faster on our hardware.

2.5 Wave Effects: Diffraction and Interference

There were two main problems created by to the wave properties of the light. First, the periodic arrangement of the DMD formed a 2-D diffraction grating, creating multiple copies of the DMD image to be formed. This problem proved easy to solve, since the zeroth-order image was clearly brighter and easily identifiable. A converging lens in the presynaptic optics produced clearly separated diffraction images. An iris excluded all images but the zeroth order.

The second problem was created by the diffraction of the light from the individual rectangular source areas on the DMD. These are rectangular apertures which create a diffraction pattern of light, some of which spills outside of the rectangular image on the film and on the detector. This was the major source of crosstalk between synaptic areas. Even a fairly small amount of crosstalk was highly detrimental to the network accuracy. We solved this problem by increasing the dark borders between the source areas to about 20% of the size of the rectangle. This diffraction imposes a limit on the number of synapses that can be handled by this optical neural network design. A goal of our future designs is to eliminate this problem.

2.6 Clock Issues: Cycles, Phases and Pulses

All neurons in our networks were synchronous. Each neuron computes its next state based on the current activations of its source neurons, but does not change its output until all neurons have finished computing their next state. Then all neurons on the same class change their state simultaneously. There are two classes of neurons: *per-pulse-update* and *per-phase-update*.

Our optical hardware network has three levels of timing. External cycle, internal phase and signal pulses. External cycles represent one network input vector being processed to generate a network output vector. That is, one *exemplar* is processed. The network input vector is applied at the beginning of a cycle and remains unchanged until the next cycle. The network's output vector is decided at the end of a cycle.

An internal phase is the time it takes a signal to forward-propagate one layer. Typically, our networks may have three to six internal phases for each external cycle. The per-phase-update neurons change their outputs at the end of an internal phase.

For pulse-based intensity modulation schemes, each internal phase is divided into a number of pulses. The number of pulses depends on the number of significant bits required for the neural signals. It was typically 256 pulses for 8 significant bits. The DMD is updated and a new CCD image acquired each pulse. The per-pulse-update neurons change their output at the end of each pulse.

3 Fixed-Weight Learning Neural Networks

3.1 The Fixed-Weight Learning Theorem

Fixed-Weight Learning (also called Adaptive or Accommodative Neural Networks) has been investigated by several researchers [4-9] However, this is (as far as we know) the first reporting of results from FWL-NNs implemented in special hardware, whether optical or electronic. Previous papers have been mostly concerned with their highly adaptive nature and/or their use in optimizing learning.

In [4] Cotter and Conwell proved the Fixed-Weight Learning Theorem: Given a neural network topology (which learns by changing weights) and its attendant learning algorithm, there exists an equivalent FWL-NN. Any mapping that can be learned by changing the weights of the original network can be learned by the FWL-NN without changing any synaptic weights.

The FWL-NN learns because a learning algorithm is encoded in its (fixed) synaptic weights. The learned function mapping information is dynamically stored in recurrent neural signals. We call these signals *potencies* (also known as *flying weights* [11]) to distinguish them from the standard synaptic weights.

A FWL-NN can learn the full range of mappings that its non-fixed-weight equivalent network can learn. However, there are costs associated with fixed-weight learning. The FWL-NN will (almost always) be larger than the equivalent changing-weight network. This is because it also has to perform the learning computations along with the mapping computations of the equivalent network. Also, FWL-NNs are necessarily recurrent even if the initial equivalent network was not.

All of the FWL-NNs presented here perform *on-line* learning. The target value of the previously presented exemplar (during the last external cycle, *t-1*) is provided to the network. Alternatively, the error of the network output for the previous exemplar could have been provided. In general, on-line learning is not a requirement for FWL-NNs.

3.2 Creating the Fixed-Weight Learning Networks

The method we used for this work assumes that the network can be divided into two main parts: the *planapse* and the *tranapse*. The planapse (from $\pi\lambda\alpha\nu\eta$ meaning *error*) performs the potency update calculations, and the tranapse performs 'the potency signal times the input signal' calculations. In our networks, there is one planapse and one tranapse for each synapse in the equivalent non-fixed-weight equivalent network.

The planapse and tranapse computations are performed by sub-networks that were trained separately and integrated together to form the FWL-NN.

Planapse. The planapse sub-network was trained to learn well-known (on-line) Back-propagation Learning Rule:

$$\Delta P(t) = x(t-1) \times y(t-1) \times (1 - y(t-1)) \times (y(t-1) - T(t-1)), \quad \text{where}$$

t	:current exemplar (external cycle)
$t-1$	:exemplar one cycle previous
x	:input to synapse
y	:output of neuron
T	:Target value for neuron
ΔP	:Change in Potency (flying weight signal)

(1)

This can be a bit confusing since the Backpropagation learning rule was used to train the planapse on the Backpropagation learning rule. Of course, other learning rules could be used for either.

The training data sets were generated by choosing random values for the inputs, and using the mapping formula to compute the targets. Note that a feedback signal consisting of the target value $T(t-1)$ for the previous data exemplar must be provided to the FWL-NN. Alternatively, an error signal, such as $e(t) = y(t-1) - T(t-1)$ could have been used as the feedback. The ΔP can be either positive or negative (bipolar), but optical intensity signals are unipolar. We scaled the calculations so that zero was represented by light as half intensity, the most negative signal was represented as no light intensity, and the most positive signal was represented by full light intensity.

Tranapse. The tranapse sub-network was trained to perform a scaled version of $s(t) = P(t) \times x(t)$. That is, a *Potency signal* times an *input signal*. We named this sub-network a *tranapse*, because a tranapse is to a (artificial) synapse as a transistor is to a resistor. The same bipolar scaling method was used as with the planapse. In addition, the tranapse output was scaled to effectively extend the potency range to $-\omega \leq P \leq +\omega$, where ω was usually 4.0.

3.3 From Software Synaptic Weights to Optical Attenuations

The (fixed) synaptic weights ranged from -10 to +10. However, attenuation of light physically ranges from 0 to 1. The sign of the synaptic weight was known by our software neurons, so the attenuation needed to only encode the magnitude of the weights. The maximum synaptic weight magnitude was determined for each neuron. Each of the neuron's synaptic weights was divided by this maximum weight magnitude to compute the required attenuation. To compensate for this weight scaling, each neuron has a constant multiplicative scale factor which is equal to the maximum weight magnitude. This "extra" scale factor should require no extra opto-electronic hardware, since the optical signal must be amplified anyway as part of the light detection process.

We tried various ways of dividing the planapse/tranapse pairs into sub-networks. For instance, should the multiplications be separately trained? Should the planapse and tranapse operations be done within the same-subnet with simultaneously training? Each of these sub-network configurations had its own strengths and weaknesses.

One lesson we learned is that the precise numerical addition of two signals that are very different in magnitude (such as the weight update from t-1 to t) is very difficult for a network to learn, and easiest done by "hand" – that is, wiring up a linear neuron with appropriate synaptic weights.

4 Experimental Testing Results

4.1 Sub-network Training

Table I illustrates (software-based simulation) performance data for a sub-network trained to perform unsigned multiplication (for instance, it may be used to perform the $x(t-1) \times y(t-1)$ calculation in the planapse). Training was performed using the MATLAB *nntool.m*, using the automatic step size adjustment option *traingdx*. We trained each of the networks for 100,000 epochs of 10,000 randomly-generated training pairs. The large number of epochs was necessary to reduce the network errors. The relatively large training data set helped reduce overlearning. All of the planapse/tranapse schemes required a similar amount of training.

We believe that the small increase in error for more than 7 hidden neurons could be reduced by more epochs training on these larger networks.

A separately generated data set was used to test the sub-networks after they were trained.

The MSE and SigBits columns were calculated from:

$$MSE = \frac{1}{N} \sum_{n=1}^{N} (y_n - T_n)^2$$

$$SigBits = -\log_2 \left(\frac{1}{N} \sum_{n=1}^{N} |y_n - T_n| \right), \text{ where} \tag{2}$$

$$N = \text{Number of Exemplars in Test Set}$$

As the table shows, the accuracy of the sub-network depends on the number of hidden nodes. This points out a property of FWL-NNs -- the *size* of the neural network required to learn a mapping set depends on the *accuracy* needed to learn the mapping set.

4.2 Testing on Optical Hardware

Table 2 shows the results of testing two neural networks on the optical hardware. The first network is an unsigned multiplication (*uMULT*). This is the same feed-forward network shown in Table 1 and Figure 2. The test data was a set of exemplars with two random inputs $x_1, x_2 \in \{x | 0 \le x < 1 \subset \Re\}$ and one target $T = x_1 \times x_2$.

The second network is *PlanTran*, a FWL-NN that is equivalent to a network with a single changing synaptic weight, with logsig squashing function and trained by Backpropagation. This FWL-NN is made from a single Planapse – Tranapse pair.

Table 1. (Simulated) Unsigned Multiplication. Hidden Layer Size vs. Mean Squared Error (MSE), and number of significant bits of result. Size of Training Set: 10,000. Epochs: 100,000.

Hidden Layer	MSE	Sig Bits
3	6.5003×10^{-4}	5.3
4	3.6876×10^{-4}	5.7
5	3.0794×10^{-4}	5.8
6	3.1636×10^{-5}	7.5
7	2.1617×10^{-5}	7.7
8	4.0069×10^{-5}	7.3
9	5.4367×10^{-5}	7.1

Generating Test Data for FWL-NNs. The algorithm to generate training/test data for a FWL-NN is:

```
repeat Number-of-Mappings times
   Randomly select a mapping M from a set of mappings S.
   repeat Number-of-Exemplars-per-Mapping times
      Generate a random input vector x
      Use x with mapping M generate target vector T
      Output training pair (x,T)
   end repeat
end repeat
```

For *PlanTran*, **S** was the set of all function mappings $T = \text{logsig}(M \cdot x)$, $-4 \leq M \leq +4$, where the real index M specifies the particular mapping. The set **S** represents all mappings that a single-synapse neural network (with logsig squashing function) can (in theory) learn exactly.

There are several important parameters that were measured during the network testing. Size of the network, the number of internal clock phases per external clock cycle (exemplars), the average number of cycles the network required to learn a mapping, the residual error of the FWL-NN after learning has occurred.

For *PlanTran*, the number of SigBits was computed by:

$$SigBits = -\log_2 \left(\frac{1}{N-L} \sum_{n=L+1}^{N} |y_n - T_n| \right), \text{ where}$$

N = Number of-Exemplars

L = Number of Exemplars Required to Learn Mapping

Table 2. Experimental Results on Optical Hardware. L- Number of Layers, N–number of neurons, W–number of synapses, ϕ – Phases per Exemplar, Pulses – Number of pulse timeslots in one Phase. Learn – Number of Exemplars required to learn mapping (for FWL-NN) , MSE – mean squared error (after learning), SigBits – Number of Significant Bits.

NN	L	N	W	ϕ	Pulses	Learn	MSE	SigBits
uMULT	3	13	30	2	128	n/a	0.0013	~6
PlanTran	4	29	100	6	256	11	0.0083	~4

5 Conclusion and Future Work

The initial *uMULT* results show that the optical hardware can perform the unsigned operation to moderate precision (six bits or more). The *PlanTran* network results demonstrates that Fixed-Weight Learning can work on an optical hardware platforms. However, a more accurate and reproducible method of creating optical attenuation than 35mm film may be necessary for larger networks.

Both of the above neural networks are fundamental "building blocks" on which larger FWL-NNs can be constructed. We are currently performing ongoing measurements and testing to extend these results.

Based on what we have learned while designing, building, and testing this optical hardware, we are designing a new hardware platform to support these new FWL-NNs. The goal of this research and development is to create a practical hardware platform capable of performing large, complex real-world neural computation tasks at very high speeds.

Acknowledgments. This material is based on work supported by the United States National Science Foundation under Grant No. 0725867.

References

1. Keller, P.E., Gmitro, A.F.: Operational Parameters of an Opto-Electronic Neural Network Employing Fixed-Planar Holographic Interconnects. World Congress on Neural Networks (1993)
2. Abu-Mostafa, Y.S., Psaltis, D.: Optical Neural Computers Scientific American (March 1987)
3. Kubler, C., Ehrke, H., Huber, R., Lopez, R., Halabica, A., Haglund, R.F., Leiterstorfer, A.: Coherent Structural Dynamics and Electronic Correlations during an Ultrafast Insulator-to-Matal Phase Transition in VO_2. Physical Review Letters. PRL 99, 116401 14 (September 2007)
4. Cotter, N.E., Conwell, P.R.: Learning algorithms and fixed dynamics. In: Proceedings of the International Conference on Neural Networks 1991, vol. I, pp. 799–804. IEEE, Los Alamitos (1991)
5. Feldkamp, L.A., Prokhorov, D.V., Feldkamp, T.: Conditioned Adaptive Behavior from Kalman Filter Trained Recurrent Networks. IEEE, Los Alamitos (2003)
6. Younger, A.S., Conwell, P.R., Cotter, N.E.: Fixed-Weight On-Line Learning. IEEE Transactions on Neural Networks 10(2), 272–283 (1999)

7. Prokhorov, D.V., Feldkamp, L.A., Tyukin, I.Y.: Adaptive Behavior with Fixed Weights in RNN: An Overview. In: IJCNN 2002. IEEE, Los Alamitos (2002)
8. Lo, J.T., Bassu, D.: Adaptive vs. Accomodative Neural Networks for Adaptive System Identification. In: IJCNN 2001, IEEE, Los Alamitos (2001)
9. Hochreiter, S., Younger, A.S., Conwell, P.R.: Learning To Learn Using Gradient Descent. In: Proceedings of the International Conference on Artificial Neural Networks, Springer, Heidelberg (2001)
10. Bade, S.L., Hutchings, B.L.: FPGA-Based Stochastic Neural Networks. In: Implementation IEEE FPGAs for Custom Computing Machines Workshop, Napa, CA, pp. 189–198 (1994)
11. Werbos, P.: Private Communication (2004)

Solving NP-Complete Problems with Delayed Signals: An Overview of Current Research Directions

Mihai Oltean and Oana Muntean

Department of Computer Science,
Faculty of Mathematics and Computer Science,
Babeş-Bolyai University, Kogălniceanu 1,
Cluj-Napoca, 400084, Romania
{moltean,oanamuntean}@cs.ubbcluj.ro
http://www.cs.ubbcluj.ro/~moltean/optical

Abstract. In this paper we summarize the existing principles for building unconventional computing devices that involve delayed signals for
encoding solutions to NP-complete problems. We are interested in the
following aspects: the properties of the signal, the operations performed
within the devices, some components required for the physical implementation, precision required for correctly reading the solution and the
decrease in the signal's strength. Six problems have been solved so far
by using the above enumerated principles: Hamiltonian path, travelling
salesman, bounded and unbounded subset sum, Diophantine equations
and exact cover. For the hardware implementation several types of signals can be used: light, electric power, sound, electro-magnetic etc.

Keywords: unconventional computing, signal-based computing, NP-
complete, delay lines, optical computing.

1 Introduction

NP-complete problems [6] have attracted a great number of researchers due to
their simple terms but huge complexity. Despite the impressive amount of work
invested in these problems no one has been able to design a polynomial-time
algorithm for them. A relatively new direction is to attack these problems with
unconventional devices. DNA computers [2], Quantum computers [5, 21], bub-
ble soap [1], membrane computing [18, 19], gear-based computer [22], adiabatic
algorithm [10] etc are few of the most important approaches of this kind.

Here we outline some of the most important principles governing some unconventional devices which use delayed signals for encoding solutions to NP-
complete problems. A common feature of all these devices is the fact that the
signals are delayed by a certain amount of time. The existence of a solution is
determined by checking whether there is at least one signal which was delayed
by a precise amount of time. If we don't find a signal at that moment it means
that the problem has no solution.

The difficulty of this approach resides in the design of a delaying system such
that the solution can simply be read at an exact moment of time.

S. Dolev, T. Haist, and M. Oltean (Eds.): OSC 2008, LNCS 5172, pp. 115–127, 2008.

At the current stage we are interested to find only if a solution exists for the investigated problem. Otherwise stated, we try to solve the decision (YES/NO) version of the problems.

Since we work with signals we need a physical structure in which the signals travel. The structure is usually represented as a directed graph with arcs connecting nodes. The directed graph is designed in such a way that all possible solutions of the problem are generated. The device has 2 special nodes: a start node (where the signal enters) and the destination node (where the signals are collected and interpreted).

Initially, the signal (pulse) is sent to the start node. As the signal traverses in graph it will be divided into more and more signals. Each of them will encode a partial solution for the problem. It is important that the signals do not annihilate each other. At the destination node we filter the solutions by checking for the good ones.

There are several other ways for solving NP-complete problems by using light and its properties. Two other different approaches have been presented in [4] and [20].

The paper is organized as follows: Section 2 describes the NP-complete problems. Properties of the signal useful for our research are described in section 3. Operations performed in our devices are described in section 4. Some examples of devices working with delayed signals are shown in section 7. Several practical aspects for hardware implementation are discussed in sections 8 - 11. Difficulties for the practical implementations are given in section 12. Further work directions are given in section 13. Section 14 concludes our paper.

2   YES/NO NP-Complete Problems

NP-complete problems [6] are a special class of problems for which we don't know whether a polynomial-time algorithm exists. There is no proof that we can solve them only in exponential time nor a polynomial algorithm was proposed so far. NP-complete problems are linked together by a polynomial time reduction. Thus, if one of them is solved in polynomial time it means that all others can be solved in polynomial-time.

NP-complete problems are usually formulated as decision problems. Instead of asking for a minimal solution (e.g. the shortest path, the smallest set, the lowest point etc) one can ask if there is a solution smaller than a fixed constant K (e.g. the length of the path is shorter than K, the number of elements in the set is smaller than K etc). These are decision problems (also known as YES/NO problems).

In our research we are interested in the solving the decision problems. We are not interested to find the actual solution of the problems.

3   Properties of the Signal That We Can Count on

Two properties of signal are of great interest for our research. Most types of signal that we know (light, sound, electric etc) have these properties.

- The speed of the signal has a limit. We can delay any signal by forcing it to pass through a cable of a certain length.
- The signal can be easily divided into multiple signals of smaller intensity/power.

For some problems it is required for the signal makes some loops (see the unbounded subset sum problem in section 7). This type of flow is not possible for the electric-based signals. This is why, when we talk about problems whose structure requires loops we assume that we work with optical signals not with electric-based.

4   What We Do with the Signals

The following manipulations of the signals are performed within the devices:

- When the signal passes through an arc it is delayed by the amount of time assigned to that arc.
- When the signal passes through a node it is divided into a number of signals equal to the out degree of that node. Every obtained signal is directed toward one of the nodes connected to the current node. In this way we add parallelism to our devices. This feature is actually the source for a major drawback: due to repetitive divisions the strength of the signal decreases exponentially and more and more powerful signals are required for larger and larger instances of the problems.

5   Basic Idea

The device has a directed graph-like structure with 2 special nodes: a start node and a destination node.

The signal is sent initially to the start node. It will then enter in the rest of the graph where the actual computations are performed.

The purpose of the destination is to collect the solutions. In the destination node we have placed a reading device which measures the moments when the signals arrive there.

In the rest of the directed graph we have nodes which split the signal and arcs which delay the signal. We work with arcs (directed edges) instead of edges (undirected edges) because we don't want to annihilate the signals coming from 2 opposite directions.

The graph must be constructed in a special way. Each signal follows a particular path meaningful for the problem structure. The signal constructs a solution by visiting nodes and arcs. When it traverses an arc it is delayed by some amount of time. Finally it arrives in the destination node. A black-box representation of our device is given in Figure 5.

If the solution of the problem was correctly constructed we will have a particular delay induced to a particular signal. In what follows we denote by B this delay. No other signals (which do not encode solutions) can have this delay.

Fig. 1. Black-Box design of our device. Signal enters from start node. Within the device, the signals follow different paths and are divided multiple times. In the destination we will get different signals at different moments of time.

This is a hard constraint. The delaying system must obey this rule, otherwise we cannot make distinction between signals representing complete solutions and signals encoding partial or incorrect solutions.

The difficulty of this approach resides in satisfying this constraint.

6   Did We Solve the Problem?

In the destination node we have more signals arriving at different moments. There can be two cases:

- If there is a signal arriving at moment B, this means that there is a solution for our problem.
- If there is a no signal arriving at moment B means that there is no solution to our problem.

If there is more than one signal arriving at the moment B in to the destination it simply means that there are multiple solutions to the problem. This is not a problem because we want to answer the YES/NO decision problem (see section 2). At this moment we are not interested in finding the actual content of the solution.

Because we work with continuous signal we cannot expect to have discrete output at the destination node. This means that arrival of the signals is notified by fluctuations in the intensity of the signal. These fluctuations will be read by some specialized device (such as an oscilloscope).

7   Designing the Graph

Figures 2, 3 and 4 show several directed graphs used for solving various problems. All graphs have a polynomial number of nodes. In what follows we describe the basic ideas behind each device.

The standard subset sum problem has the simplest design [11, 16]. Each number can appear or not in the final solution. This decision is represented in our device by 2 arcs having the same 2 nodes as extremities. One of the arcs has 0 length and the other arc delays the signal by an amount of time equal to one of the numbers from the given set. If the signal traverses the arc having the length greater than 0 it means that the corresponding number is selected in the solution. If the signal traverse the 0 length arc it means that the corresponding number is not selected in the solution. Practically we cannot have arcs of 0 length. This is why a constant k is added to all arcs. $(B + n * k)$ is the moment when the existence of a solution should be checked (where B is the target value of the problem, n is the cardinal of given set and k is a constant). Finally, since we want to sum all numbers in solution we construct the device in a serializable way (see Figure 2 a)). The delays are polynomial (in the size of the given numbers).

The design for unbounded subset sum contains fewer arcs compared to standard subset sum due to constraint relaxation (each number can appear multiple times in the solution). This is why we don't have to create a serial structure with no return arcs. Instead a loop structure was proposed [11, 12]. The internal node is used for dividing any incoming signal into $n + 1$ subsignals (where n is the cardinal of the given set). n signals are sent back to arcs encoding numbers and the $(n + 1)^{th}$ is sent to the destination (see Figure 2 b)).

The Hamiltonian problem asks to visit each city exactly once (see Figure 3 a)). It is not easy to satisfy this constraint since we cannot restrict the signal to visit a node exactly once. More than that, the distance between nodes is not important in this problem, thus if incorrectly designed can lead to multiple rays arriving in the same moment in destination. Because the constraints are imposed by nodes, the delays should be focused on nodes instead of arcs. This is different from the previously described solutions where the only purpose of nodes was to divide the signal. Let us suppose that the signal encoding the Hamiltonian path arrives at moment B in the destination. No other signals (not encoding Hamiltonian paths) must arrive in the same moment there. We have to choose the delay induced by each node in order to satisfy this constraint. In [14, 15] it was shown an example of such delaying system. That system guarantees that the delay induced to the signal encoding a Hamiltonian path will not be equal to any other path visiting some cities more than once or skipping some other cities. Unfortunately it was exponential (the length of the delays increases exponentially with the number of nodes).

The directed graph for the Exact Cover problem [6] (see Figure 3 b)) is a combination between Hamiltonian path and standard subset sum [17]. Some subsets from a collection must be chosen (like in the standard subset sum) and each number from the original set must appear exactly once (like nodes in the Hamiltonian path). The delaying system is exponential because it is based on the Hamiltonian path device.

For solutions to Diophantine equations we have to choose some positive integer numbers x and y which have the property $a_1 * x + a_2 * y = c$ [11, 13] (where a and b are some positive integer numbers). A brute-force approach was employed by generating all possible pairs (x, y). The trick consists in a loop whose purpose is to increase the value of a variable with 1 unit. The signal enters in the loop and

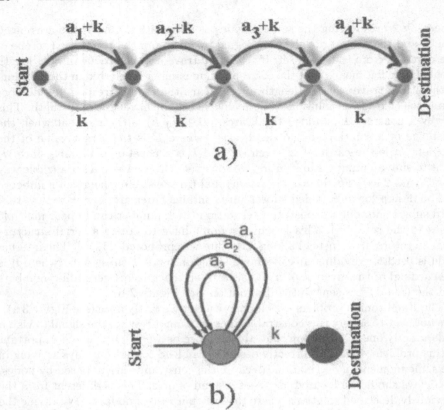

Fig. 2. The graphs representing devices for a) standard subset sum, b) unbounded subset sum. A problem with 4 numbers $\{a_1, a_2, a_3, a_4\}$ is considered for the standard subset sum. The constant k was added to all arcs because we cannot have cables of length 0. All arcs have been depicted similarly, but in reality they can have different lengths depending on the values of the numbers in the given set. A problem with 3 numbers $\{a_1, a_2, a_3\}$ is considered for the unbounded subset sum. Signals encoding combinations of numbers arrive in the internal nodes and are sent either to destination or back again for adding more delays.

traverses it. When exits it will be divided into 2 signals: one of them will be sent to the next node and another one will be sent back to the loop for increasing the delay another time-unit (see Figure 4 a)). Because we cannot have cables of length 0 we have to search for a solution at moment $c + 2 * k$, where k is the delayed induced by cables connecting the nodes. The delaying system is polynomial.

The construction of TSP device imposes a double difficulty: some nodes must be visited exactly once and the total path must be the shortest possible [7, 8]. If we ignore delays on nodes we will have paths not being Hamiltonian. If we focus only on delays induced to nodes, we will not obtain the shortest path. To

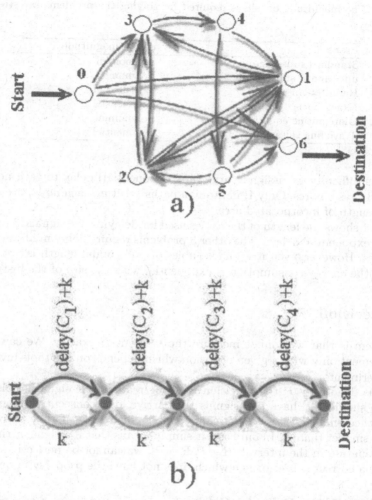

Fig. 3. The graphs representing devices for a) Hamiltonian path, b) Exact Cover. In the case of Hamiltonian path, the length of the arcs connecting nodes is not important - it can be the same for all arcs. What is important are the delays induced by each node. Thus, inside each node we have another set of arcs - not depicted here - which introduce some delays. The device for Exact Cover is more complicated. Since we wanted that each number from the original set to appear exactly once we use the delaying system from the Hamiltonian path. However, this is not enough, because here we have to select sets from a given collection and not single numbers. For this purpose we have assigned to each set a delay equal to sum of delays attached to numbers from that set. It is denoted by $delay(C_i)$. Choosing the correct set of sets is done in a similar manner with the standard subset sum (see Figure 2). Note that numbers from original set cannot be seen in this picture. They are actually hidden inside the delays induced by each set.

Table 1. The magnitude of delays required for solving the problems investigated in this paper

Problem	delays magnitude
Standard subset sum	polynomial
unbounded subset sum	polynomial
Hamiltonian path	exponential
Exact Cover	exponential
Diophantine equations	polynomial
Travelling salesman	exponential

solve this difficulty we assign a large (and exponential) delay to each node and smaller delays on arcs. Only Hamiltonian paths (visiting each node) are checked for the length of incorporated arcs.

Table 1 shows the length of the cables used for delaying the signals. 3 problems requires exponential delays. The other 3 problems require polinomial time length for cables. However, even if we have cables of polynomial length is not enough because the energy consumption is exponential with the size of the instance.

8   Precision

A problem is that we cannot measure the moment B exactly. We can do this measurement only with a given precision which depends on the tools involved in the experiments.

Let us denote by P the precision used for reading our signals. This means that we should not have two signals that arrive at 2 consecutive moments at a difference smaller than P. We cannot distinguish them if they arrive in an interval smaller than P. In our case, it simply means that if a signal arrives in to the destination in the interval $[B - P, B + P]$, we cannot be perfectly sure that we have a correct subset or one which does not have the property in question.

8.1   What If We Delay by Cables ?

Let us denote by v the speed of the signal. Based on that we can easily compute the minimal cable length that should be traversed by the signal in order (for the latter) to be delayed with P seconds. This is obviously $v * P$ meters. This value is the minimal delay that should be introduced by an arc. Assuming a $3 * 10^8 m/s$ for the optical signal and a 10^{-12} precision of the best oscilloscope we get a $3 * 10^{-4} m$ for the shortest cable that we can have in our system.

This value is the minimal delay that should be introduced by an arc in order to ensure that the difference between the moments when two consecutive signals arrive at the destination node is greater than or equal to the measurable unit of P seconds. This will also ensure that we will be able to correctly identify whether the signal has arrived in to the destination node at a moment equal to the sum of delays introduced by each arc.

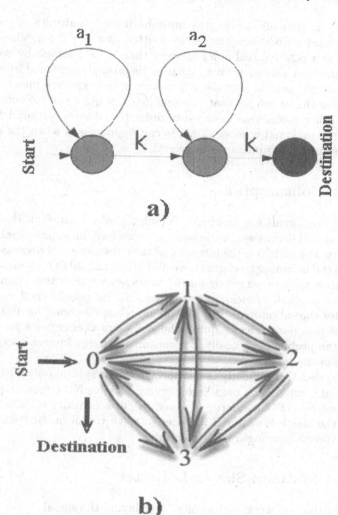

Fig. 4. The graphs representing devices for: a) Diophantine equations, b) travelling salesman. A loop of the device for Diophantine equations is similar to the device for unbounded subset sum problem with only one number in the set. Signals looping through device are actually increasing values for x and y. Note that the device can be extended for any number of variables. Travelling salesman is again a complicated device. Practically $n!$ Hamiltonian paths can be generated and we are interested in searching for the shortest one. First step is to ensure that we can distinguish between Hamiltonian and non-Hamiltonian paths. This is done as it was explained in Figure 2. To check for the shortest path we add some delays for each arc connecting nodes. These delays must be significantly shorter than the delays within nodes so that the discovery of Hamiltonian paths is not affected.

A constraint is that all the lengths must be integer multiples of $v * P$. We cannot accept cables whose lengths can be written as $x * v * P + y$, where x is an integer and y is a positive real number lower than $v * P$ because by combining this kind of numbers we can have a signal in the above mentioned interval but that signal does not encode a subset whose sum is the expected one.

Once we have the length for that minimal delay, is quite easy to compute the length of the other cables that are used in order to induce a certain delay.

Note that the maximal number of nodes can be increased when the precision of our measurement instruments is increased.

9   Energy Consumption

Within nodes the signals are divided into (sub)signals. Because of that, the intensity of the signal decreases. In the worst case we have an exponential decrease of the intensity. For instance, the intensity of the subsignals will decrease k times (compared to the incoming strength) if we divide each signal in k subsignals. If we do this operation n times we get signals k^n times weaker than the original signal. Even if we have a small branching factor (the smallest possible is 2 - utilized in the solution for the subset sum [16]) we still get a huge decrease for 100 nodes.

This means that our devices require a huge amount of energy for solving large instances of the problems. Actually, the consumed energy increases exponential with the size of the instance.

Please note that this difficulty is not specific to our system only. Other major unconventional computation paradigms, trying to solve NP-complete problems, share the same fate. For instance, as noted in [9], a quantity of DNA equal to the mass of the Earth is required in order to solve Hamiltonian Path Problem instances of 200 cities using a DNA computer.

10   Speed Matters: Slower Is Better

Assume again that we work with cables for delaying the signal.

The speed of the signal is an important parameter in our device. Working with a high speed signal is bad for our device due to the precision problems exposed in section 8. We can either increase the precision of our measurement tools or decrease the speed of signal.

By reducing the speed of signal by 7 orders of magnitude, we can reduce the size of the involved cables by a similar order (assuming that the precision of the measurement tools is still the same). This can help us solve larger instances of the problem.

11   Basic Components for Physical Implementation

For implementing the proposed device we need the following components:

- a source of signal (laser, pulse generator etc),
- Several splitters for dividing a signal into multiple subsignals. If we work with electric signals the split is trivial. If we work with light we need some beam-splitters (such as half-silvered mirrors).
- A device for reading the fluctuations in the signal intensity. If we work with electric signals we need an oscilloscope. If we work with optical signals we need either a combination of a photodiode and an oscilloscope or a special device for reading optical signals. Another possibility for optical signals is to use white light interferometry [7].
- A set of cables used for connecting nodes and for delaying the signals.

11.1 How to Introduce Delays ?

There are several ways in which the signals can be delayed. These variants depend on the type of signal to be delayed.

- delay lines (optical or electrical). Electric delays are induced by either long lines o by discrete inductors and capacitors [23].
- columns of mercury (for delaying sound waves). These devices have been originally used as memory in old computers.

12 Difficulties

Several difficulties might be encountered during the construction of such devices. Some of them are listed below:

- Building a general purpose device able to solve a wide range of problems and instances. This is a critical aspect for making these devices practical. Further issues are discussed in section 13,
- Setting the delays to an exact value. If we work with cables we have to cut them with huge precision. Electrical delay lines have a non-zero rise time which can introduce further difficulties to the system,
- Providing enough power to the system in order to be able to solve large instances of the problems. This is the greatest difficulty for our device and cannot be solved unless P = NP,
- Finding high precision reading instruments. Due to high speed of the signals we need very good reading instruments for detecting very small and very fast fluctuations in the intensity of the signal.

13 Automation

Currently the design and construction of graphs for each problem is made by hand. This dramatically reduced the area of applicability. Automating the process of building the devices would represent a huge step for practical applications. For achieving this purpose we have to use/design the followings:

- a scalable and reconfigurable graph. This should allow us to enable/disable arcs between nodes. The graph should be large enough to accommodate various sizes of the problems.
- several programmable / reconfigurable delay lines. In this way we can easily modify the delay quantity induced by each arc. Electrical delay lines with up to 256 steps are already available on the market, which means that we can easily have 256 possible values for delays. By serializing such devices we can have larger ranges of values.

14   Conclusions

The way in which signal can be used for performing useful computations has been investigated in this paper. The techniques are based on 2 properties of the signals: the massive parallelism and the limited speed.

Several important aspects have been exposed in this survey: what kind of operations are performed with the signals, how to construct the graph for several problems, how to find if the problem was solved or not, how to cope with precision and power decrease, which are the basic components required for implementation and which are the most common difficulties encountered during the physical implementations.

By using the described methods several problems have been solved so far: Hamiltonian path, travelling salesman, bounded and unbounded subset sum, Diophantine equations and exact cover.

Future works directions are focused on: implementing the presented devices, solving new problems and automating the construction process.

References

[1] Aaronson, S.: NP-complete problems and physical reality. ACM SIGACT News Complexity Theory Column, March. ECCC TR05-026, quant-ph/0502072 (2005)
[2] Adleman, L.: Molecular computation of solutions to combinatorial problems. Science 266, 1021–1024 (1994)
[3] Bajcsy, M., Zibrov, A.S., Lukin, M.D.: Stationary pulses of light in an atomic medium. Nature 426, 638–641 (2003)
[4] Collings, N., Sumi, R., Weible, K.J., Acklin, B., Xue, W.: The use of optical hardware to find good solutions of the travelling salesman problem (TSP). In: Proc. SPIE, vol. 1806, pp. 637–641 (1993)
[5] Feynman, R.: Simulating physics with computers. International Journal of Theoretical Physics 21, 467 (1982)
[6] Garey, M.R., Johnson, D.S.: Computers and intractability: A guide to NP-Completeness. Freeman & Co, San Francisco (1979)
[7] Haist, T., Osten, W.: An Optical Solution For The Traveling Salesman Problem. Opt. Express 15, 10473–10482 (2007)
[8] Haist, T., Osten, W.: An Optical Solution For The Traveling Salesman Problem:erratum. Opt. Express 15, 12627–12627 (2007)
[9] Hartmanis, J.: On the weight of computations. Bulletin of the EATCS 55, 136–138 (1995)

[10] Kieu, T.D.: Quantum algorithm for Hilbert's tenth problem. Intl. Journal of Theoretical Physics 42, 1461–1478 (2003)

[11] Muntean, O.: Optical Solutions for NP-complete problems, graduation thesis, Faculty of Mathematics and Computer Science, Babes-Bolyai University, Cluj-Napoca, Romania, defended 3rd of (July 2007)

[12] Muntean, O., Oltean, M.: Using light for solving the unbounded subset-sum problem (submitted, 2008)

[13] Muntean, O., Oltean, O.: Deciding whether a linear Diophantine equation has solutions by using a light-based device (submitted, 2008)

[14] Oltean, M.: A light-based device for solving the Hamiltonian path problem. In: Calude, C., et al. (eds.) UC 2006. LNCS, vol. 4135, pp. 217–227. Springer, Heidelberg (2006)

[15] Oltean, M.: Solving the Hamiltonian path problem with a light-based computer. Natural Computing 7(1), 57–70 (2008)

[16] Oltean, M., Muntean, O.: Solving the subset-sum problem with a light-based device. Natural Computing (in press, 2008)

[17] Oltean, M., Muntean, O.: Exact Cover with light. New Generation Computing 26(4) (2008)

[18] Paun, Gh.: Computing with membranes. Journal of Computer and System Sciences 61(1), 108–143 (2000)

[19] Paun, Gh.: P systems with active membranes: attacking NP-complete problems. Journal of Automata, Languages and Combinatorics 6(1), 75–90 (2001)

[20] Shaked, N.T., Messika, S., Dolev, S., Rosen, J.: Optical solution for bounded NP-complete problems. Applied Optics 46, 711–724 (2007)

[21] Shor, P.W.: Algorithms for quantum computation: Discrete logarithms and factoring. In: Goldwasser, S. (ed.) Proc. 35nd Annual Symposium on Foundations of Computer Science, pp. 124–134. IEEE Computer Society Press, Los Alamitos (1994)

[22] Vergis, A., Steiglitz, K., Dickinson, B.: The complexity of analog computation. Mathematics and Computers in Simulation 28, 91–113 (1986)

[23] Delay line memory @ Wikipedia (accessed) (12.06.2008),
http://en.wikipedia.org/wiki/Delay_line_memory

Author Index

Lecture Notes in Computer Science

Sublibrary 1: Theoretical Computer Science and General Issues

For information about Vols. 1–4921
please contact your bookseller or Springer